THE THIRD
INDUSTRIAL
REVOLUTION

THE THIRD
INDUSTRIAL
REVOLUTION

by G. Harry Stine

G. P. Putnam's Sons, New York

To
ARTHUR C. CLARKE

"Why get excited about anything that doesn't go straight up?"
—Personal remark to the author,
White Sands, New Mexico,
August, 1952.

SBN: 399-11552-8
Library of Congress Catalog Card Number: 75-21783

PRINTED IN THE UNITED STATES OF AMERICA

Contents

Introduction

EYE-OPENING testimony about space processing and manufacturing was received in the Senate in 1973. Before the Senate Aeronautical and Space Sciences Committee, David Keller of the General Electric Company announced that he had identified markets for space processing and manufacturing that could have an annual value of over $2 billion in the 1980s.

Keller stated that he had found more than one hundred processes that might be produced or manufactured in space and for which a demand exists. He selected only ten areas that met the criteria of market demand, technical and economic feasibility, and gave these estimates of their dollar values: (1) ten typical vaccines—over $1 billion; (2) livestock sperm—$422 million; (3) DC rectification and regulation crystals—$100 million; (4) tungsten carbide components for oil pumps and valves—$60 to $85 million; (5) aircraft turbine blades—$36 million; (6) acoustic wave devices—$100 million; (7) X-ray targets—$18 million; (8) glasses (scientific and optical)—$10 million; (9) computer memory devices—$380 million; and (10) development of small electric motors—$56 million.

Stated another way: commercial markets for space manufacturing and processing over the ten-year life cycle of the Space Shuttle could amount to as much as $20 billion. Looking to the future, a second and third generation of Space Shuttle could drastically reduce the cost of putting payloads into orbit, thereby opening up new commercial applications.

I believe it is important to remember that Keller's findings

7

were made at a time when the potential of the Space Shuttle system was just beginning to be understood. Historically, transportation systems have tended to greatly exceed the expectations of their inventors, both technically and economically. Imagine the difficulty that anyone would have had in 1492 predicting the size and value of transatlantic traffic in the twentieth century based on the voyages of Christopher Columbus.

Skylab showed that man can survive in space for long periods of time, and demonstrated that he could do work in a weightless state that cannot be done on Earth. Large crystals and purer alloys are examples.

Against this background, Harry Stine has written a thought-provoking book that is an imaginary extension of the Keller testimony and of the *Skylab* experience.

The Third Industrial Revolution comes from an author of note who has had a distinguished career in aerospace engineering and has a demonstrated capacity for looking into the future intelligently. Space, he argues, is to be a scene of a Third Industrial Revolution because there man can find virtually limitless energy and resources. Pollution as a by-product of the First and Second Industrial Revolutions disappears in the vastness of space.

He pictures our present earthbound industrial system as being a closed system for ecological purposes. By developing space as a site for industry, man opens up the system and insures his future survival—a survival holding the promise of plenty rather than scarcity.

While any forward-looking book can be attacked because it is visionary, let us remember another space "visionary," Dr. Wernher von Braun who once stated that all progress is based on dreams.

The Third Industrial Revolution reflects a happy combination of hope and optimism, which are vital ingredients of America's greatness, and it provides a real challenge to the gloom and doom crowd.

Phoenix, Arizona　　　　　　　　　　BARRY GOLDWATER
November, 1974

THE THIRD
INDUSTRIAL
REVOLUTION

CHAPTER

1

The Coming Revolution

A FAR-REACHING and final revolution is going to take place in our lifetimes. This revolution has already started.

It is going to change the face of Planet Earth. It will drastically alter our life-styles. It will affect nearly every person on Earth. It holds the promise of improving the quality of life and of increasing the standard of living of those who wish to participate.

Out of this revolution, new industrial empires will be forged. New billionaire industrial moguls will emerge. Thousands of new products will be created—some of them impossible to make today and others prohibitively expensive with today's technology.

During this revolution, we will begin to transform Planet Earth into a garden planet.

Every revolution holds forth these or similar promises. Is this all pie in the sky? No, but the sky is where it will all happen, starting one hundred miles over our heads in space.

It is the Third Industrial Revolution, the final industrial revolution, the exploitation of the Solar System. It will be brought to fruition by practical, hardheaded, "bottom-line" people.

We can only begin to sense the full, encompassing impact of the industrial utilization of space. To venture the forecast that space-based industrial operations may, in the next hundred years, play a major role in our planetary destiny may seem at this time to be too highly speculative. But there is a

rationale to the forecast that is based upon long-term historical trends.

In the field of astronautics, we are still in the early exploratory phases of development, with the exception of certain
types of unmanned satellites—communications satellites and
meteorological satellites, for example. In any area of human
activity and accomplishment, progress begins with *exploration*. The technologies involved are always very crude, very
primitive, and very expensive. Exploration is always carried
out with high risk to human life; explorers expect this. Exploration is nearly always financed by governmental or quasi-
governmental organizations.

Pragmatic, rational, hard-working, conservative, and
hardheaded people usually consider exploration to be a
waste of time and money. This is because exploration produces only information. It consumes rather than produces
capital. But without exploration, there is nothing to exploit
in the future. Exploration, research, and development keep
the kitchen cabinets of knowledge well stocked so that when
the entrepreneur comes to the cabinet in search of new information to give him a competitive advantage, he will find
something there. The trick is for the explorer to know when
to stop asking for surplus funds from his patron, and for the
patron to know when to cut back the funds to keep the explorer from consuming too much capital. Ever since the
scientists and engineers got together with the financiers,
merchants, and manufacturers about two hundred years
ago, this delicate balance has always been a bone of conten-
tion. By and large, the balance has worked out well. For example, David Sarnoff made a highly successful operation of
RCA by striking a workable balance between support of research and development and profitable manufacturing and
marketing.

Exploration must be followed by exploitation if the exploration is to be meaningful and relevant. Research, development, and investigation must be followed by use of the information produced or the capital expended in exploration
is wasted. Knowledge for the sake of knowledge is in itself a

dead end. The quest for knowledge is a dead end without an overall long-term goal of utility. The history of exploration/ exploitation cycles shows this clearly. A gross example of the failure of exploration is the Norse discovery of America circa A.D. 1000; they could not exploit their discoveries, and the western hemisphere had to await the energies of Spain that were awakened by the conquest of the Iberian Peninsula from the Moslems. Sir William Crookes made thorough investigations of glowing green phosphors inside electron tubes, but the television tube that grew from this exploration lay unused and unneeded on the laboratory shelf for over half a century. (Who knows what scientific and technical wonders now lie on the laboratory shelf?)

Once exploration becomes exploitation, the explorer is no longer required. In fact, the explorer is actually discouraged. If he has any sense, he becomes the "elder statesman" of the field and basks in well-earned recognition, perhaps sweetened with a small piece of the action.

Then the center of the stage is occupied by the entrepreneur. He is the capitalist because exploitation generates more capital than it consumes. It creates wealth. It is no longer the province of government bureaucracies, but the realm of free enterprise. The government that financed the original exploration does well when it backs off and recoups its investment by taxing the entrepreneur. By managing material and human resources efficiently, parsimoniously, and compassionately, the entrepreneur creates something of value that other people are willing to exchange value tokens for. If he does his job right, the entrepreneur ends up with a little more than what went into the system to start with. And so does everyone else.

Admittedly, this is a rather simplistic consideration of a very complex system. But, perhaps by paring it back to such basics, we can use it as part of the overall rationale to back up the "atrocious and wildly speculative" statements that opened this book.

And it all adds up to the logic of the Third Industrial Revolution.

The Third Industrial Revolution is astronautics passing from the stage of exploration to the stage of exploitation. The first glimmerings of this transition are already evident. The Third Industrial Revolution is already beginning to happen amid studies and data purporting to show that we are running out of natural resources, that we are rapidly depleting energy sources, and that we are polluting and devastating our planetary ecology.

In a broader view, these "counterindustrial" contentions are only part of what appears to be a deepening worldwide megacrisis.

This megacrisis has many facets. It's an energy crisis. It's a population explosion. It's a matter of increasing levels of pollution, growing food shortages, several local ecological imbalances, and a depletion of known reserves of natural resources. It has been the subject of numerous serious studies of the "where-are-we-going?" type. The most notorious of these was the study, *The Limits to Growth*, by D. H. Meadows, *et al.*, for the Club of Rome. The Meadows study considered the Earth as a closed system, a cage in which all of us are trapped forever, a planet-sized terrarium in which we and our progeny are fated to spend our days forevermore. When the megacrisis is considered as occurring within a closed system, the human race does indeed appear to face a future of pestilence, famine, destruction, decadence, shortages, and the New Dark Ages.

"There is a major problem involved with falling back into a New Dark Age," says Dr. Krafft Ehricke, a pioneer futurist. "If we slip back into the Dark Ages this time, we do so with our fingers on the nuclear trigger."

With this extremely sobering thought in mind, we should take a careful look at the possible courses the megacrisis can take. There appear to be three possibilities:

1. The current course of affairs will continue with all problems worsening until catastrophe occurs, the nuclear trigger is stroked, and the human race destroys itself and the ecology of the Planet Earth in the process.

2. The planetary ecology reacts to what we are doing to it

and rebalances the state of affairs . . . at the expense of the human race.

3. The human race—all of us—gets busy and rebalances the planetary system in such a way that the solution arrived at allows both the human race and the planet to continue their historical development and evolution.

The first possibility may be tagged the "catastrophic future." It is the one referred to by Ehricke. However, the nuclear trigger amounts to a One-Second Slum Clearance Program; it will probably do far more long-range harm to the human race than to the planet. Those who believe that we can truly destroy the Planet Earth have more faith in the power of technology than the technologists themselves. The Planet Earth is a damned big place . . . even at six hundred miles per hour in a jet airliner. From the standpoint of probability, this catastrophic future is quite unlikely because we are neither lemmings bent on self-destruction nor omnipotent planet-busters. Believers in this possibility lack perspective. It is truly possible; but it is also the most unlikely direction.

The second possibility may be labeled "Malthusian." It is more likely than the previous choice, mainly because we are still very stupid when it comes to a true understanding of the highly complex and interrelated subject of ecology. We don't yet understand all of the synergistic relationships. In our ignorance there is a reaction of fear. The unlikely aspect of this possibility lies in the fact that we have started to heed the warning flags that have started to fly.

Since the beginning of the twentieth century, we have undertaken massive conservation and reclamation projects. If you will look at old photos taken in the American West in the 1890's you will be surprised at the sparse forests where dense stands of timber exist today. The second possibility has a low probability because we have not only started to think about solving the problems, but we have already spent decades doing something about it.

The third possibility is the only rational one. It is the only choice that we, as human beings, can possibly abide. It is not "living with the problem"; that is an animal response to the challenge. It is solving the problem for our benefit and survival as well as solving the problem for our home planet, on which a couple billion of our progeny are going to have to live at any one time, regardless of whether some of the human race eventually expands outward in the great Stellar Diaspora of the next millennium. (Not all of our own ancestors left the eastern hemisphere to find new lives in the New World.)

The Third Industrial Revolution is only *part* of solving the problem under the terms of the third choice.

There is a further rationale to this third possible future that lends very strong credence to the concept of the Third Industrial Revolution. This involves an insight that comes from considering the studies of people whose profession is the study of mankind: the anthropologists. They are interested in the long-term view. They are superhistorians who consider the development of the human race in terms of millennia rather than years or even centuries.

Anthropologist Carleton S. Coon, in his book *The Story of Man,* discusses three Newtonian-like biological/sociological laws that seem to be recognized and understood only by the life scientists. These three laws are universal:

1. *The Law of Evolution:* Once an organism or organization of organisms has evolved into a successful, surviving entity, it will not undergo further evolutionary changes to its physical characteristics for a very long time.

2. *The Law of Least Effort:* Any organism or organization of organisms will opt for the survival solution that requires the least effort to achieve and sustain.

3. *The Law of Acceleration:* Biological parameters always exhibit an exponential, cumulative increase up to the moment when a physical climax occurs; this climax may be caused by direct physical limitations or by damping from other factors.

Coon, viewing history from the anthropologist's long-range vantage point, also proposes a most intriguing thesis:

"Man has been converting energy into social structure at an ever-increasing pace. As he has drawn more and more energy from the earth's storehouse, he has organized himself into institutions of increasing size and complexity."

We have achieved today's earth-girdling community of nations and people only because we have the energy resources to do it. An incredible level of energy expenditure is required to sustain the economic, technological, and cultural level of most of the world's population. The relevance of Coon's energy thesis can be sensed by viewing any large city such as New York, London, or Tokyo from the air at night. If we are going to do what we are talking about here—solving the megacrisis problems by the third possibility—we will have to find the energy (and the materials) to do it and sustain it. We cannot maintain a worldwide community of mankind on the energy of muscle power alone.

Very well. Here we are talking about solving the world's megacrisis and insuring our future survival as a life form. And yet we are running out of energy, raw materials, land area, etc. In the closed system of Planet Earth, we already know that we are limited in energy sources alone.

If, according to Coon, we are indeed engaged in the business of converting natural energy into social structure, and if all our social institutions require increasing levels of energy consumption if they are to expand to solve the problem, and if the resources of the closed system of Planet Earth are to be inadequate, one answer alone remains:

Neil Armstrong knew exactly what it was when he spoke of the "giant leap for mankind."

We must open up the system as rapidly as feasible and profitable.

We must come to the realization that the Solar System and all that is in it is *not* simply a place full of information for us to ferret out and contemplate. We must view the Solar System as a useful place for human purposes . . . in fact, a *necessary* place for us to live, work, and use.

As Dr. Krafft Ehricke puts it, "The world is no more closed than it is flat."

Or, to use the blunt words of Robert A. Heinlein, "We've just about used up this planet; time to go find another one."

This is the philosophy behind the concept of the Third Industrial Revolution. Every worthwhile goal in human affairs has been accompanied by its philosophy. Those transient achievements without an understandable supporting philosophy die on the vine. Witness the Apollo Program. The monumental achievement of landing men on another celestial body paled almost at once upon its attainment of its goal because its spokesmen could not voice any acceptable philosophy to accompany it. Therefore, it appeared to be a stunt. The philosophers did not understand the achievement, in spite of the fact that thousands upon thousands of people were deeply committed to the program and could, at least privately, form their own personal philosophies about it. But dwelling upon failures does not produce progress, which is what we are discussing in this book.

Some very astute scientific minds outside the aerospace fraternity have already glimpsed the concept of opening the system by making our way of living into a three-dimensional world, by expanding it into the Solar System. Princeton physicist Dr. Gerard O'Neill, for example, has discussed the potentialities of colonies of human beings living in artificial worlds in space; the paper was published in a journal no less prestigious than *Physics Today*. Pioneer futurist Dandridge M. Cole of General Electric Company wrote of this in 1961. But one essential element was missing from all these forecasts.

People will not voluntarily leave Mother Earth and travel to live in or on a foreign world just to escape Earth, no matter how difficult the problems grow to be here. Many of our forefathers left Europe not only because conditions were bad there, but also because there was promise in America; there was something to do here; *there was a job to be had and work to be done.*

And the same is and will be true of living elsewhere in the Solar System. Those space stations or colonies or bases must offer something to do. Not only must the Solar System be useful, but it must offer something to be done.

And it will, because the Solar System is the most logical choice of all sites for mankind's growing industrial operations.

In the century to come, we will have to move most of our industrial operations off Planet Earth and reestablish them in the Solar System where they will not and cannot disturb any ecological systems. There will be some industrial operations that will remain on Planet Earth because they do not pose any hazard to the terrestrial biosphere; this is Krafft Ehricke's "benign industrial revolution."

But the Third Industrial Revolution is the one that is already starting to take place in the Solar System.

The term "industrial revolution" is a very specific one that describes what we have been doing since the early eighteenth century in Europe and America . . . and more recently all over the world. A *revolution* is a rapid and radical change in the way of doing things, of social organization, of life-style. *Industry* is the sum total of human activities involved in the process of converting natural energy into social structure in accordance with Coon's hypothesis. Therefore, an industrial revolution is a drastic change in the work operations, products, and manual-mental output of human beings.

Actually, we have been in an industrial *evolution* (which is a *gradual* change) from the time people first made tools (and weapons) using fire. From the simple days when proto-humans fire-hardened sharp points on wood shafts for hunting to today's worldwide industrial complex, people have required energy of increasing intensity as well as the use of materials and processes of higher and higher levels of complexity, difficulty, and energy consumption.

Why is the shift to space industry called the Third Industrial Revolution?

The First Industrial Revolution began in England in the eighteenth century. It reached its peak in Europe and America between 1875 and 1910; it has yet to reach its peak in several parts of the world, while other areas have yet to progress beyond it. It introduced a number of new ways of doing things:

It introduced powered machinery to replace human and animal muscle power, which is a strictly limited source of energy.

It featured the conversion of fossil fuel reserves to heat energy by exothermic chemical combustion processes; this heat energy was then converted to mechanical motion that propelled or operated the machines.

It resulted in the development of mass-produced devices made with interchangeable parts on a production line of people who carried out one and only one assembly operation per person per device assembled.

Its legacy includes air and water pollution, the dehumanization of industrial operations, the demise of cottage industry, the disappearance of the craftsman, and a host of other highly undesirable consequences that need not be elaborated here; see any environmentalist.

How and why did it start? Two factors may have triggered the First Industrial Revolution. First, England may have found herself stripped of manpower that was needed overseas to administer and defend her growing colonial empire, which was needed to feed her growing industries, and so on around the circle. Second, England found herself beset by an energy shortage equivalent, from the point of view of time, to today's energy crunch. The last of England's hardwood forests had fallen to the woodsman's ax. Now it has often been said that England is the only place in the world where it is possible to freeze to death atop one of the world's greatest coal deposits. The shortage of wood forced the English to figure out a way to get coal out of deep mines that were continually flooded by water. They either had to develop the technology to pump the water out of those deep coal mines . . . or freeze to death. Getting the water out required pumps, and the pumps required energy. Iron was required to harness the energy that came from expanding steam. To get the iron to build the Newcomen atmospheric engines and the later Watt steam engines required high temperatures. Formerly, these high temperatures were obtained from charcoal obtained from the fine hardwood forests.

It is possible to trace the development of all modern industry and its technological base back to a single common ancestor, one triggering invention. This was Abraham Darby's coke oven, perfected in the early 1700's. This permitted the sulfur to be extracted from the English soft coal to produce

coke that would smelt iron that would build the steam engines that would power the pumps that would remove the
water from the mines so that the coal could be brought out
and burned or put into Darby's coke oven to make more
iron. Industrial operations began to feed upon themselves,
to improve themselves, to lift themselves by their own boot-
straps. Today, a Watt steam engine, as seen in museums—
and they are impressive devices in operation—seems to be incredibly crude, but it was the very best technology available
at the time. Someday the Space Shuttle will appear as crude
to our progeny.

Within one hundred and fifty years, the entire industrial
world that we know today grew from practically nothing. Before it had run its course, this revolution forever changed the
life-style of Europe, Japan, and the North American conti-
nent. In time, it changed every part of the Planet Earth. Today, we find it very difficult—if not impossible—to imagine
what it was like before the First Industrial Revolution. Many
people who talk about a "return to the simple life without
modern technology" apparently hate technology so much
that they have never bothered to study it and therefore don't
know what it would be like without it. We have come so far in
the last two hundred years that most of us don't consciously
grasp the total magnitude of the change that has taken place.

Yes, the First Industrial Revolution did have its expensive
consequences. We are still living with many of them today,
and we will have to continue living with some of them for a
few years yet; we cannot afford the luxury of throwing away
these processes because they are at the bottom of the industrial pyramid and we don't yet have better or more efficient
methods with which to replace them.

One of the sociological implications of the First Industrial
Revolution was its complete destruction of the previous
agricultural base of living, wherein one family, working as
hard as it could, provided the food for one other family not
on a farm, as well as for itself.

A big question that could be debated for years: Could it
have happened any other way?

The Second Industrial Revolution began in the United

States of America early in the twentieth century, although there were some early glimmerings of it in Europe before that time. This revolution involved the use of feedback and logic devices to replace the human brain—and therefore the human being—in the repetitive or very rapid tasks of machine operation or direction. It occurred because the devices and processes developed in the First Industrial Revolution matured to the point where a human being could not react quickly enough to exercise adequate and timely control. Or perhaps the human could not handle the complex interrelationships of some of these advanced processes. For example, a petrochemical refinery has a lot of processes going on all at the same time. Some of them happen very quickly. Others require the proper adjustment of Items A, B, C, and D all at the same time. A human being just can't do it for an hour, let alone eight hours at a stretch. Furthermore, he doesn't do a very good job of it at all. There are too many things to do. He's worse than a juggler with ten balls in the air at once. So the technologists developed the fly-ball governor, the automatic stop, the limit switch, the fuse, the overload switch, the Jacquard loom, the servomechanism, magnetic amplifiers, feedback loops, and all the other devices of the Second Industrial Revolution. While the snorting, throbbing, pounding steam engine with its "monkey motion" rods and valves (and clouds of black smoke) is representative of the First Industrial Revolution and the nineteenth century, the Second Industrial Revolution and the machinery of the twentieth century turn out to be the silent, winking computer with its spinning tape decks and chattering readouts.

The Second Industrial Revolution didn't eliminate people. But it required people who could use their brains as well as their hands—simply because people told the machines what to do and fixed the machines when they broke. There is a classic J. R. Williams "Bull of the Woods" cartoon in which the automatic machine operator is complaining, "But the automatic stop didn't stop!" To which the shop foreman replies, "Well, why weren't you watching the automatic stop, hah?"

The Second Industrial Revolution produced another complete change in our way of life between the early 1900's and today. Some of us can still remember hand-stoking a home coal furnace before the automatic Iron Fireman was installed. There was a time when all elevators had a human elevator operator, when all telephones were reached only through a human operator manning a switchboard at "central."

As with the First Industrial Revolution, once the Second Industrial Revolution got rolling, many new industrial processes became possible, practical, economical, and profitable (in that order). Greater amounts of energy were required, even though the industrial processes became more economical; there were more processes requiring more energy. This accelerated the depletion of the fossil fuel deposits. It also combined with another factor to create some real trouble.

The increased complexity of industrial processes resulted in some very complex industrial waste products—and these were *really* waste products because there are no natural ecological recycling mechanisms for them. Pollution itself became a problem because of this and because of the increased number of industrial operations. There are *always* waste products. Get any engineer to tell you about a thing called the Second Law of Thermodynamics, a fancy name for a natural law that says there are always waste products from anything that is done.

The industrial operations resulting from the First and Second Industrial Revolutions on Planet Earth have now grown to such a size, social cost, and complexity that it is very difficult—if not impossible, and exceedingly expensive (the increasing cost of industrial products bears witness to this)—to expand these operations, improve them, eliminate all or most of their polluting capacity, or make innovations within the suddenly perceived closed-cycle ecology of this small planet.

The Second Industrial Revolution is reaching its peak because the size of this planet is finite.

But, despair not, for the first two Industrial Revolutions

have paved the way for the third one. They have permitted us to develop the technology, crude as it is at this moment, to climb well out of our planetary gravity and to live in the environment of the Solar System. We can do it because we *have* done it. Yes, the technologies involved are crude, clumsy, inefficient, expensive, wasteful of natural resources, and otherwise uneconomical . . . *at this moment!* This will not long be the case with space-related technologies. All early, emergent technologies exhibit these frightening characteristics. But they become highly refined, efficient, economical, and very commonplace in a surprisingly short period of time *if they are useful.*

And space-related technologies are useful because they are the stepping-stones to the future. They'll mature in less than one hundred years. This forecast can be justified by the biological Law of Acceleration, which also applies to information, learning curves, and time required to complete a development cycle.

We should put this forecast into perspective in order to justify it. History is the only perspective we have to go by. Approximately one hundred and fifty years were required to go from Darby's coke oven, at the start of the First Industrial Revolution, to Lee DeForest's triode vacuum tube that heralded the start of the Second Industrial Revolution. But a mere sixty-five years separated the triode from the pioneer experiments in space manufacturing—*Apollo-8* and *Soyuz-6.*

It will not require one hundred and fifty years—or even sixty-five years—to mature the Third Industrial Revolution.

This is partly because information is synergistic—a word borrowed from biology meaning that when two or more items work together, they produce an effect that neither is capable of alone. The male-female combination is a perfect example of this. In the performing arts, the synergism of music and electronics has produced almost an entirely new art form. In addition to being synergistic, information is also accumulated on the exponential curve of the Law of Acceleration, and the buildup becomes startling very quickly. Meadows, *et al.*, made a beautiful case for it in their celebrated study of Closed System Earth, *The Limits to Growth.* We are currently doubling the quantity of accumulated knowledge once every seven years. In essence, the size of the local public library is doubling in less than a decade; just look at the books in that library, and it will give you some inkling of what is taking place with information. The application of this knowledge by engineers, with the financial backing of entrepreneurs and the management talents of executives—all of which are driven by the desires of the marketplace—have resulted in many new industrial processes and products. All of this has required greater energy inputs. This, in turn, has further accelerated the use of fossil fuels and raw materials.

Remember, the most difficult task imaginable is making the very first perfectly flat steel plate to make the first machine lathe to make the first stamping press. After you build the first of these basic machine tools, you can use them to make more machine tools. The most expensive and valuable thing in the world is the first prototype of a new device, be it a machine, a computer program, a drawing, or a manuscript. It is the pattern from which all duplicates are made and from which will spring all of its progeny.

We do *not* have to reinvent the wheel to make the Third Industrial Revolution happen. We have nearly all the knowledge. We have nearly all the machines required. There are no new scientific breakthroughs required. What remains to be done to bring the Third Industrial Revolution to maturity is the long, gut-tearing, nitty-gritty fiddling with pesky figures and drawings and sketches and prototypes and equipment to work out the bugs. There is a lot of work to be done.

At this moment in time, we are primarily concerned with getting into space on a large scale so that we can use the Solar System. Learning how to conduct industrial operations in space will occupy our best technical people for the next century. It will involve developing new industrial operations in orbit in the Earth-Moon system first—in near-Earth orbit, in lunar orbit, and on the Moon itself. Eventually, within fifty years, space industrial operations will grow to encompass most of the Solar System, as we will discuss herein.

Don't look at a steel mill and shake your head, wondering how we'll ever move it off the Earth and into space . . . because we won't. In the first place, it's designed to work on Earth with an atmosphere around it and gravity working on all parts of it. It won't work in space, even if we could transport it there for a penny per pound . . . which we can't do and probably won't be able to do. The steel mill will die . . . literally; its place will be taken by industrial operations of low energy consumption and low pollution potential. The "benign industrial revolution" on Earth will replace the steel mill with another kind of factory attuned to human characteristics and merged with the environment; it will be fed by the heavy industry of the Solar System whose products drop from the sky like Biblical manna. The new steel mill will be built in space with the planetary remains of the astroid belt to feed its zero-gravity blast furnaces powered by solar energy; there is no biosphere or planetary ecology to pollute out there, and it will be a long time before this minuscule interruption of the energy flow of the universe amounts to polluting the universe . . . and there are answers to that, too.

The Solar System is probably the best possible place to locate industrial operations, freeing the Planet Earth for people, giving us the chance to return the planetary ecology to something like the way it was one hundred thousand years ago when we were biologically attuned to it.

We will then have what will truly be a garden planet, and we will also have the heavy industrial base to support an advanced and high-level living standard.

And this alone will make the Third Industrial Revolution the final industrial revolution.

CHAPTER

2

Industry: What Is It and Who Needs It?

SITUATED AS WE are in the opening years of the Third Industrial Revolution, the whole affair and the forecasts about it may appear to be highly incongruous to many people. The tasks remaining to be done and the problems yet to be solved may appear at first glance to be overwhelming, if not impossible. All forecasters and planners face this problem when they begin to deal with any projection that extends more than ten years into the future . . . and especially with any forecast that involves extensions and developments of technology, even though the technological base exists. A very recent example of this can be cited. At the end of World War II, the jet airplane was a new device, full of development bugs, surrounded by emergent technologies, and requiring the best possible people to operate it. To suggest that a little old lady would be able to fly around the world in armchair comfort aboard a jet plane at nearly the speed of sound within fifteen years was an unbelievable forecast. It turned out fifteen years later that little old ladies could indeed do this; within twenty-five years, it was the *only* way a grandmother could visit her grandchildren because the jet airplane spelled the demise of the transatlantic ocean liner and most of the long-distance passenger trains. The capital investment to create the international jet airliner network was staggering to the mind; it involved not only the aircraft, but whole new airports, incredibly complex new airways control systems, complicated navigational systems, fuel storage and distribution systems to feed the voracious thirst of the jet engines, ground transportation systems, maintenance organizations, food handling operations, and the entire chain of systems and subsystems to support these.

To have suggested this possibility in 1910 would have seemed insanity. In forecasting something as complex and long-term as the Third Industrial Revolution, one runs up against the same "crisis of believability," even though all the elements are there, waiting to be activated by the world's men of action.

Skepticism comes easy to most Americans because it is part of our culture and developed historically from the earliest days of the American colonies. We are a pragmatic people, with little use for speculation other than as entertainment. Yet we are willing and eager to accept a grand challenge if it has a goal, real or imagined, of bettering our lives or those of our children.

Many of us were also trained in a way of thinking that can be labeled either the "scientific method" or "engineering school." Or it may have been called the "college of business administration." This training discourages speculation in public, particularly with the resourses of others. Although it is perfectly acceptable in the privacy of one's own laboratory or office to engage in the wildest flights of fancy beyond any possible hard proof by theory or numbers, one cannot and must not engage in such things in public.

In addition,"systems thinking" has begun to spread from the engineering sciences into the world of business. One now tries to consider all the myriad details possible in an endeavor in order to discover critical paths, suitable options, the best balance, workable compromises, and possible tradeoffs and their consequences.

Laid against this background of thinking processes used by the people who will make the Third Industrial Revolution happen—the entrepreneurs, risk-takers, bankers, managers, and executives—the mere thought of establishing an industrial base in the solar system as the Earth-based system winds down, becomes obsolete, or succumbs to environmental pressures is indeed an overwhelming consideration. Today's problems are taxing enough; why worry about pie in the sky?

Happily, the world still contains a lot of people of the same sort who built the railroads, the steel mills, the petroleum industry, the airlines, and all the other great entrepreneurial risks of their time. These are the people who look beyond the bottom line. They are the ones who build new industrial empires. They are the ones who will also become the space moguls. Some of them have already seen the coming of the Third Industrial Revolution, although many of them may not have thought of it in that term. Some of them are already involved, turning the crank to make it happen.

Like earlier times during the First and Second Industrial Revolutions, the space industrials are today strongly opposed by the neo-Luddites.

In 1811 as the First Industrial Revolution was just getting up steam, organized bands of rioters appeared in the English countryside around Nottingham; they were intent on destroying the new textile-manufacturing machinery because they believed it would eliminate jobs for weavers and other craftsmen. These Luddites took their name from the mythical King Ludd or Ned Ludd. The relative prosperity that began to grow from the First Industrial Revolution finally broke up the Luddites.

They exist today in a new guise as the neo-Luddites. They don't destroy machinery or industrial plants. Through the communications media, whose workings they know very well, the neo-Luddites attempt to destroy industrial operations before they become reality through attempts to manipulate public opinion.

The Third Industrial Revolution is going to become a prime target for the neo-Luddites. We already hear such statements as, "Why bother? We are entering a postindustrial culture. Industry is no longer important! We must stop developing new products and new ways of doing things . . . but *you must* figure out some way to stop polluting. The main thrust of human activity from now on must be in the development of better human relationships and greater social institutions. Industry is passé. Since we don't need it anymore, why even bother to think about having it in space?"

"Industry" was defined earlier as the sum total of human activities involved in the process of converting natural energy into social structure.

Quite literally, our society would collapse without industry. Most of the people in the world would die within a year without our industrial base. If anyone thinks that we can get along with less of an industrial base, he should visit those parts of this planet that still don't have such a base and learn what it is like to try to live without it. He should ask those people who are working very hard to develop their local industrial base; they'll certainly tell him of the hardships involved and of the life-style without industry. Or he may wish to consult the ghosts of his ancestors who worked to develop the current industrial base and who did without the luxuries he considers to be necessities today.

Far too many people on Planet Earth live without an industrial base to supply their basic survival needs. Their only method of survival without industry has been ingrained in their culture by untold centuries: Take it by force from those who may have it.

An industrial base means that people do not have to take those necessities from someone else; *they can make them instead.*

The only human relationship possible in a society devoid of modern industry is that of master-slave and haves-versus-have-nots.

Who needs industry? Everyone in the world needs industry . . . and the more of it the better.

And the only way to get more of it is to do it where it won't further harm our planet.

Simplistic? Perhaps. But perhaps we need simplistic approaches and definitions in a culture of growing complexities.

Since the Third Industrial Industrial Revolution is such a complex concept, we can get a better understanding of it by breaking it down into simpler elements. This is a problem-solving technique that is widely used and is so common that most problem-solvers really don't think too much about it. A big problem always becomes more manageable when it is broken down into smaller, simpler elements. We can then concentrate on those elements and, eventually, put the whole works together down the line as a complete system.

(There are many similar problem-solving techniques that are widely used and that have grown from this basic one. You may be familiar with some of them and call them by names such as PERT. If you are a professional problem-solver—an engineer or a manager—bear with us. *You*, too, may discover that the Third Industrial Revolution isn't the impossible task you might have thought!)

The Third Industrial Revolution, as we will see in depth later, involves many different industrial operations, processes, and end products. Let us first consider *any* industrial operation, analyze it down to its simple elements, and then look at the space-based industrial operation's problems . . . or lack of them, as the case may be.

What is an industrial operation? What are its characteristics, regardless of what it is, where it is located, or what is its end product?

Any industrial processing operation has the following elements—some operations have more or less than others, but all of them have these elements in varying degrees:

Raw materials to process into finished products.

Energy to make the industrial process go, and the energy is sometimes incorporated into or locked up in the final product.

Equipment to manipulate the raw materials, to apply the energy, and for creating:

An environment for the raw materials to be processed through; this may be high vacuum, high temperature, high pressure, any combination of these, or their inverses, for example.

A location for all of this to exist in.

Personnel to oversee, manage, start up, and maintain the industrial operation.

A *heat sink,* which is a place to discharge the waste energy that cannot be used by the operation; this usually comes out as heat, but may take other energetic forms as well.

A *garbage disposal system* to take care of the waste material of the industrial process.

A *warehouse,* one or several, in which to store the raw materials until they are used in the process and the finished products after they are manufactured.

A *market* consisting of a real or imagined desire to possess or use the finished product by a varying number of people.

A *finished product,* which is a device, material, or service with a market that is the *raison d'être* for the industrial operation.

A *transportation system* to convey the raw materials, finished products, and personnel between the raw material source, the location of the operation, and the location of the market.

Any industrial operation has all of these elements. Some industries, by their nature, have some in greater proportion than others. But they apply equally to all—from the simplest cottage industry to supranational industrial conglomerates to a child's sidewalk lemonade stand, a steel mill, or space industry. They apply regardless of the economic system standing behind them—capitalistic free enterprise, socialistic, or communistic (although not all of those economic systems cause the industrial operation to work equally well).

Someone is certain to think at this point that the most important element of all has been forgotten: money.

Money can be looked upon as the way to keep score in the game. Back to Square One in economics, a dollar (pound, shilling, franc, yen, lira, or rasbuknik) is basically a token for keeping score. It doesn't even have to be real; it can exist as an imaginary thing only on paper; it can exist in the future as credit.

Money has many of the attributes of energy as an industrial operations element.

So it hasn't been forgotten.

A lot of it is going to be required, and a lot more of it is going to be created by the Third Industrial Revolution. Money is the reason for doing the Third Industrial Revolution at all, and it will be a result of the Third Industrial Revolution, but only *one* of the results because of the fact that it is basically the way we keep score.

Now that we have so neatly dissected an industrial operation into twelve elements, what problems do we have to solve concerning each element to make the Third Industrial Revolution a going affair? What do we need in each area that we don't already have? What progress must be made in each area before the Third Industrial Revolution becomes both a technical and an economic reality? Is the Solar System a suitable place to conduct an industrial operation? What industrial operations could be carried out there? Can they be conducted there better than on Planet Earth? Are there any new industrial operations that we can conceive of today that can be conducted only in space?

Let us look at each element individually, but not necessarily in the order given above, keeping in mind the basic philosophical background discussed previously.

CHAPTER

3

Getting There

AT THE PRESENT moment in time and for a few years yet to come—at least until the early years of the 1980 decade—the hovering specter of high space transportation costs will continue to occupy center stage for the Third Industrial Revolution. A space transportation system is the key element in the real start-up of the Third Industrial Revolution. Until we are able to claw our way up out of Earth's persistent and strong gravity well into the Solar System with the space-going equivalent of the DC-3 airliner of the 1930's, the Third Industrial Revolution may remain a mere academic discussion. Therefore, it is vitally important to discuss the subject of space transportation at the onset. This is the one area where lack of information on true costs or projected costs may cause the vital free-enterprise capitalist to dismiss the entire Third Industrial Revolution as an interesting possibility for his great-grandchildren to consider. And it is the one area that will draw the most concentrated attacks from the neo-Luddites, who would be happier if most people would "sit at home and watch their TV sets the way God intended them to."

A considerable amount of time, money, and effort have been devoted to the subject of space transportation. Very little of this has had any real effect upon the reality of space transportation.

Since 1961 we have been putting people into artillery shells and shooting them into space. This is not a space transportation system, even if you put wings on it and call it a space plane. Look at the history of flight into space, and you will discover the reason.

Modern space rocketry began in Germany. Rockets were small devices capable of lifting perhaps a few pounds into the Earth's stratosphere until, in the early 1930 decade, General Dr. Ing. Walter Dornberger bluntly told Dr. Wernher von Braun and his engineering colleagues that they wouldn't stand a chance of getting any money to build a space rocket large enough to do any good unless they had a military requirement to back it up. So Dornberger verbally laid down the military specifications for a rocket that would carry ten times the payload of the World War I *Paris Gun* to twice the range of the *Paris Gun* with less than half the dispersion of that long-barreled weapon. Furthermore, the rocket had to be able to go through any railroad tunnel in Europe while mounted horizontally on a flat car.

The outcome of these specifications was the famous German A.4, otherwise known as the V-2 or *Vergeltungswaffe-Zwei*. It was a rocket-powered artillery shell, and it nicely circumvented the long-range artillery prohibitions of the Treaty of Versailles because rockets weren't mentioned in said treaty—only cannon.

After World War II, both the USA and the Soviet Union launched captured A.4 rockets as high-altitude research vehicles. Both nations also improved on the A.4. The liquid rocket propulsion system of the A.4 was the point of departure for the development of the Soviet rocket motors that propelled the *SS-6* and *SL-4* space-launch vehicles that lofted *Sputnik* and *Vostok*. On the American side of the world, the A.4 became the Redstone, designed by the same team that developed the A.4. The Redstone artillery missile became the *Mercury-Redstone* that lofted Alan Shepard and led to the propulsion system of both the *Atlas* and *Titan ICBM* long-range artillery rockets that launched American astronauts into orbit. Even the mighty *Saturn 5* moon rocket has a direct line of development traceable back to the German A.4, the rocket-powered artillery shell.

All of our space-launch vehicles in the mid-1970 decade

are based upon military developments for long-range artillery shells. They are designed and operated with the same philosophy that is used by an artillerist: The payload is the only important item, and the vehicle is expendable.

It has often been said that an orbital trip in a rocket vehicle—to say nothing of a lunar voyage—is analogous to making an ocean crossing in the RMS *Queen Elizabeth 2* and scuttling the ship upon arrival.

As a result, this artillery-based rocket-powered shell philosophy for space vehicles made the initial costs of putting a pound of payload in orbit about $5,000 per pound for *Vanguard* and *Explorer I,* our initial unmanned space satellites. As the research and development costs of these expendable vehicles were written off through successive launches of throwaway boosters, this cost came down to about $500 per pound of payload delivered into a 200-kilometer near-Earth orbit. In some cases, using the big *Saturn* superboosters, the cost can go as low as $150 per pound. But you have to have a lot of payload to take advantage of this discount tariff.

We do not have a true space-transportation system at the moment. This is because we have stuffed payloads into existing vehicles. When it comes to manned systems, we have literally shot people into space in artillery shells.

The United States government, through the National Aeronautics and Space Administration, is currently establishing itself as a government monopoly in the space transportation business with a program called Space Shuttle. This is not necessarily a criticism of the Space Shuttle program or the people involved in it. It is a statement of fact, regardless of the buzz words used to describe it.

Space Shuttle is probably the only way to go at this moment. It is an initial approach to a space transportation system. It suffers from a number of political compromises made in the early 1970s, trade-offs that severely curtailed the possibility that it could be an economical space-transportation system.

Space Shuttle has been described in great technical detail elsewhere, and NASA will happily tell anyone all about it just for the asking. Since it is truly the first real departure from the rocket-powered artillery shell concept, it embodies sound engineering principles, including the one that says, "Don't try too many new things all at once." The solid propellant booster rockets are discarded into the ocean; hopefully, they will be recovered and reused. The big liquid propellant tank is jettisoned in Earth orbit; it reenters the atmosphere and burns up. What does come back is the Orbiter with its payload, crew, airframe, and propulsion units intact and ready for another flight with minimum refurbishment.

With the Space Shuttle, we're approaching a true space transportation system comparable to an airline system. At least, we'll get the Orbiter back.

Current estimates by government officials put the price of a pound of payload into a 200-kilometer near-Earth orbit with the Space Shuttle at a mere $140.

This is going to be the tariff, in 1974 dollars, during the period 1980 to 1985.

But Space Shuttle isn't the end-all, any more than *Vanguard* was the first and only satellite launcher. Space Shuttle is going to breed its own progeny. In the first place, once the Orbiter is flying, history tells us that it is extremely likely that NASA will go back and ask for money to develop a completely reusable liquid-fuel booster for the Orbiter, thus creating a completely reusable system. This is what they really wanted in the first place, but they didn't think they could get the money for all of it.

Using a manned, fly-back, reusable booster vehicle for the Space Shuttle Orbiter, orbital costs in the last half of the 1980 decade could drop to as low as $38 per pound.

This cost really doesn't look too bad when we compare it with some other, more familiar costs of transporting people.

The amount of energy required to put a pound of payload into a 200-kilometer-high near-Earth orbit is about the same as that required to send a pound of payload by subsonic jet airplane halfway around the world—New York City to Sydney, Australia, for example. At the time of this writing it costs about $750 to send a person by air from JFK to SYD.

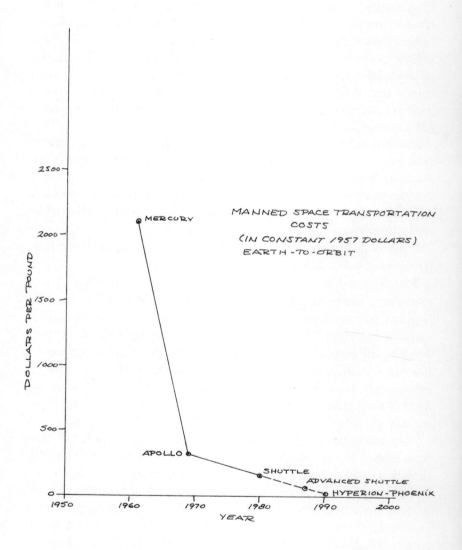

MANNED SPACE TRANSPORTATION
COSTS
(IN CONSTANT 1957 DOLLARS)
EARTH-TO-ORBIT

For the average 170-pound passenger, this works out to about $4.41 per pound.

This is about the cheapest way to do it, however. But because more people travel by car than by plane, if it were possible for a 170-pound person to drive the 12,500 miles by car at the rate of 18 cents per mile, the cost would become $13.23 per pound.

Even at a cost of only three times that much, it will be feasible to send people into orbit for industrial purposes. Today, innumerable companies have people "on the road" as salesmen, resident engineers, technical representatives, etc. This cost is staggering when computed in the cost-per-pound criterion that is usually applied to astronautics. Assume that our 170-pound man is on the road 250 days per year and travels 25,000 miles by car at a cost of $25 per day for expenses plus 18 cents per mile for the car. These are quite conservative figures. The cost works out to be $63.23 per pound. Transportation costs alone in this example work out to $26.47 per pound.

The orbital cost of $38 per pound no longer sounds quite so outrageous.

And in the years to come, the cost will come down in spite of inflation because of the historical trends for all forms of transportation. There is no reason why space transportation should not be subject to the same engineering and economic laws that have governed other transportation systems in the past and have resulted in considerable cost reductions.

For example, the domestic airline passenger fare in 1936 was 11 cents per mile in 1936 dollars (or 32 cents per mile in inflated 1972 dollars). In 1972 the domestic airline passenger fare had dropped to 6.5 cents per mile in 1972 dollars.

Similar reductions have taken place in both automobile and train travel.

There are other reasons why the $150 per pound Earth-to-orbit cost figure of the Space Shuttle should not be taken seriously by potential space moguls when laying out the cost sheets for projected space enterprises.

Space Shuttle is a government project; it is still part of the

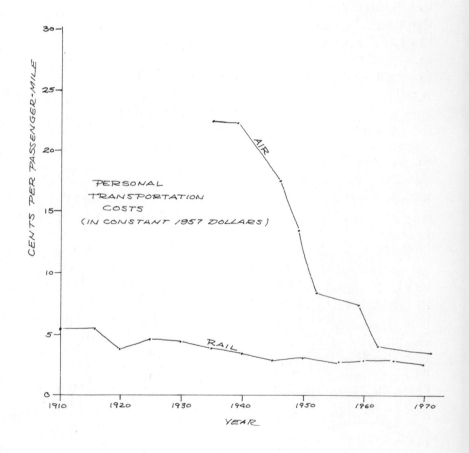

exploratory phase of space industry, as we discussed earlier. It is a necessary step toward the Third Industrial Revolution and must be allowed to continue toward operational status, primarily because it will prove beyond a shadow of a doubt that the forecasts and prognostications of this book are realistic and feasible. There is a joke among engineers, however, that says, "An elephant is a mouse built to government specifications." Space Shuttle is a government-funded program, as it must be since free enterprise must be convinced of the profitability of space industry. But it is built to government specifications that, quite rightly, were originally intended to insure that the taxpayers got what their tax dollars were paying for. Space Shuttle is costing a great deal of money because it is a government-funded program and because it is the first of its kind.

There have been several attempts historically to come up with a privately funded Earth-to-orbit ship, and there are several current attempts being made to do this.

In 1967 and 1968 Philip Bono of McDonnell Douglas— builders of the DC-3, the DC-8, the DC-9, and the DC-10, among others, and therefore one of the outstanding companies in the world when it comes to building profitable transportation devices—presented numerous technical papers before such august bodies as the New York Academy of Sciences, for example. Basing his design philosophy on the fact that an airline that buys a DC-9 doesn't fly it once, but uses it for 50,000 flight hours or more, if possible, Bono first came up with an Earth-to-orbit vehicle called *Rhombus* that looked like no rocket vehicle ever envisioned before. Like the Space Shuttle, the *Rhombus* proposal threw away several expendable metal rocket propellant tanks. But it was capable, on paper, of carrying 400 to 500 tons to orbit at a cost of $25 per pound. (And this was in 1967!)

Bono refined his conceptual design to come up with the *Hyperion*, a single-staged rocket transport capable of carrying 48 passengers and a crew of 4. It would be propelled by liquid oxygen and liquid hydrogen. It was launched on a rocket sled up the side of a mountain. It was designed to land verti-

Artist's conception of the MacDonnell-Douglas "Hyperion" space shuttle, designed by Philip Bono, being launched from a rocket sled up the side of a mountain. "Hyperion" could theoretically deliver 10,000 pounds of payload from the Earth's surface to near-Earth orbit at a cost of $12 per pound. The "Hyperion" would be completely reuseable. *(Courtesy Phil Bono and MacDonnell-Douglas Corporation)*

REUSABLE HYPERION VTOVL ROCKET CONFIGURATION

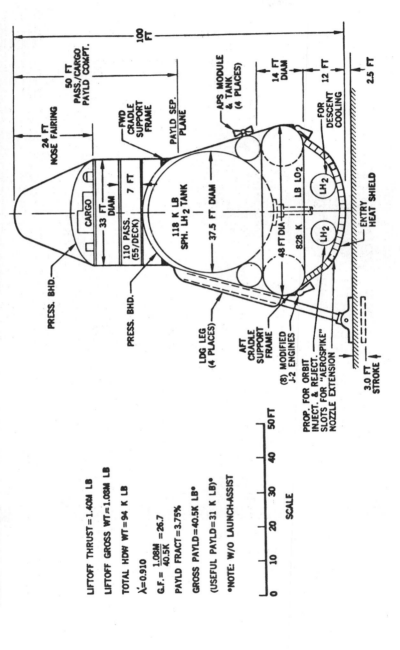

LIFTOFF THRUST = 1.4M LB

LIFTOFF GROSS WT = 1.08M LB

TOTAL HDW WT = 94 K LB

$\dot{\lambda}$ = 0.910

$G.F. = \dfrac{1.08M}{40.5K} = 26.7$

PAYLD FRACT = 3.75%

GROSS PAYLD = 40.5K LB*

(USEFUL PAYLD = 31 K LB)*

*NOTE: W/O LAUNCH-ASSIST

SCALE

0 10 20 30 40 50 FT

24 FT NOSE FAIRING

50 FT PASS./CARGO PAYLD COMPT.

100 FT

FWD CRADLE SUPPORT FRAME

PAYLD SEP. PLANE

APS MODULE & TANK (4 PLACES)

14 FT DIAM

12 FT

2.5 FT

PRESS. BHD.

CARGO

33 FT DIAM

7 FT

110 PASS. (55/DECK)

118 K LB SPH. LH₂ TANK

37.5 FT DIAM

48 FT DIA

828 K LB LO₂

LH₂

LH₂

FOR DESCENT COOLING

ENTRY HEAT SHIELD

3.0 FT STROKE

PRESS. BHD.

LDG LEG (4 PLACES)

AFT CRADLE SUPPORT FRAME

(8) MODIFIED J-2 ENGINES

PROP. FOR ORBIT INJECT. & REJECT.

SLOTS FOR "AEROSPIKE" NOZZLE EXTENSION

REUSABLE ROCKET-PROPELLED "AIR-CUSHION" SLED
(USING LAUNCH VEHICLE MOTORS)

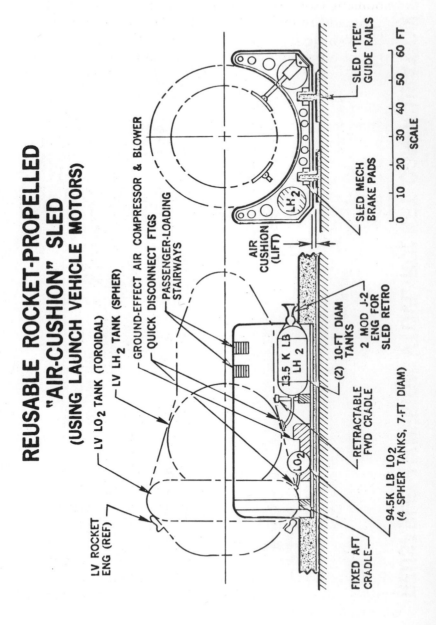

LV ROCKET ENG (REF)

LV LO$_2$ TANK (TOROIDAL)

LV LH$_2$ TANK (SPHER)

GROUND-EFFECT AIR COMPRESSOR & BLOWER

QUICK DISCONNECT FTGS

PASSENGER-LOADING STAIRWAYS

SLED "TEE" GUIDE RAILS

60 FT

50

40

SCALE

30

20

10

0

SLED MECH BRAKE PADS

AIR CUSHION (LIFT)

13.5 K LB LH$_2$

(2) 10-FT DIAM TANKS

2 MOD J-2 ENG FOR SLED RETRO

RETRACTABLE FWD CRADLE

LO$_2$

94.5K LB LO2 (4 SPHER TANKS, 7-FT DIAM)

FIXED AFT CRADLE

cally rather than glide to a runway landing like an airplane. Development costs were estimated at $2 billion. (In 1952 Boeing risked $16 million on the development of the 707 jet airliner.) Vehicle cost was estimated at $35 million, about half that of a *Concorde* SST or about equal to the cost of a Boeing 747 in 1974.

Hyperion would deliver a payload to Earth orbit for $12 per pound.

Nothing much came of these fantastically detailed design proposals, but they paved the way to show what could be done if one wished to really apply himself to the solution of the problem of high Earth-to-orbit costs.

Several prototype space entrepreneurs began in 1973 to see if they could build their own space shuttles. One of these, Space Merchants, Inc., determined that it would be possible to build an Earth-to-orbit single-staged reusable space transportation vehicle called OSIRIS—using available off-the-shelf hardware, including liquid-propellant rocket motors—for considerably less than $1 billion in development costs. The OSIRIS vehicle would be capable of delivering about 40,000 pounds of payload to a 150-mile Earth orbit at a cost of $10 per pound or less.

To gain some perspective, let's ask the question: Is this level of capital expenditure out of the question for free enterprise?

The answer is no. General Electric paid $71 million in 1973 for the 225,000-ton supertanker TT *Brooklyn*. Offshore oil-drilling rigs cost from $30 million to $50 million each. An offshore exploratory drillship can cost $50 million or more. As of mid-1974 there were 268 mobile drilling rigs for offshore petroleum drilling and 140 under construction or planned. This is a capital equipment inventory of between $12 billion and $20 billion. Boeing Airplane Company has sold over 1,000 Model 707 jet airliners at a price ranging between $4 million and $7 million each; that's between $4 billion and $7 billion worth of aircraft of a single model from a single manufacturer over a fifteen-year period.

Thus, when we are speaking about the most difficult and

expensive of the space transportation vehicles, the Earth-to-orbit ship, we are not speaking of vehicles whose development, purchase, and operational costs are in the league with the expensive, expendable *Saturn 5* moon rockets. Earth-to-orbit space transport vehicles in the 1980-1990 time period will cost no more to develop, purchase, and operate than today's large jumbo-jets, and considerably less than most oceangoing vessels and supertankers.

The Earth-to-orbit rocket transport is a device waiting for someone to exploit in order to provide access to prime industrial sites in space. It can be built with existing technology. It can be purchased within the capital expenditure capability of any number of corporations today. It can be operated at costs-per-pound equivalent to transoceanic airfreighters of today.

By 1990 industrialists and their controllers will no longer be afraid of high space transportation costs, nor of investments in space transports and industry. By that time there is likely to be a fantastic shortage of orbital weightlifting capability.

At this juncture, for those who may be sharpening their pocket calculators attempting to estimate orbital freight costs, it should be pointed out that we will not and should not use the Space Shuttle or any other manned Earth-to-orbit space transport for sending raw materials up. Any Earth-to-orbit space transport system that is probably attainable between now and the end of the twentieth century should be used to deliver to orbit the highly qualified industrial engineers and technicians or industrial research scientists and bring them back again.

As we will later see, it will be cheaper in terms of expended energy to obtain nearly all of the required raw materials from the Solar System rather than to lift them out of the Earth's deep gravity well.

It will be necessary, of course, to transport *some* raw materials and basic "seed machines" from the Earth's surface to orbit. These seed machines are those basic industrial tools and devices that will permit the construction or assembly of other devices, the basic technical equipment that will form the foundation for space industry.

Nor will manned space transports be used to deliver finished products from space *to* the surface of the Earth. Costs for the manned systems will certainly drop to mere dollars-per-pound to get something from orbit to the ground, because there is very little propellant energy expended as well as very little heat shield ablation on current and projected designs. The optimum space-to-surface industrial transportation system—a space-going boxcar, if you will—need be nothing more than an unmanned ballistic reentry body with a simple autopilot, an ablative heat shield, a recovery parachute (for those loads coming down on land or requiring very low landing shock), and a retro-rocket to nudge it into a reentry flight path. At the proper time in the flight or in the orbital path, the autopilot obeys its simple program and orders the boxcar to assume a predetermined attitude, just like one of today's unmanned interplanetary probes. The retro-rocket provides a gentle shove—no more than a few miles-per-hour change of velocity is required. Using the attitude control system, the autopilot then drops the load right on target on the Earth below, just as the *Gemini* capsules were steered toward their recovery ships by similar autopilots.

The cost for delivery of a space-going boxcar to Hong Kong, Cape Town, Le Havre, or San Francisco will not be greatly different; the autopilot merely programs a different retro-thrust vector. Ground-based terminal guidance may also be used.

It may eventually be cheaper to manufacture such space-going boxcars in space from materials obtained in space, making them expendable. There has been a lot of study devoted to putting various designs of space vehicles into space and using them as nuclei of space stations. In reality, we may well build things in space and land them on the Earth to be expended, used, and never sent back into space again!

Even transportation of materials, goods, and people from the surface of the Moon to the Earth need not be expensive. Some Earth-launched space transports will certainly make use of mountain catapults on Earth to help them into space, and there is no reason why lunar catapults cannot also be built. The Moon has no sensible atmosphere and a surface gravity one-sixth that of Earth. Its escape velocity is only 2.4 kilometers per second. Engineering designs already exist for lunar surface catapults capable of doing the job. A typical design utilizes electrical power generated from solar energy by means we will talk about in detail later. This electrical power is used to electromagnetically accelerate our space-going boxcar horizontally off the surface of the Moon to any spot on the Earth. Shaping its flight trajectory is a simple matter of celestial mechanics already well understood. The velocity at which the boxcar leaves the catapult can determine its subsequent flight path to the Earth and its eventual landing spot.

If it isn't already apparent, the true cost of transportation in space is in terms of the energy required to go from one place to another. Indeed, this even determines the cost of transportation around Earth's surface. Once out of the deep gravity wells of the planets and major satellites of the Solar System, getting around requires comparatively little energy, but often a great deal of time. If you are willing to expend more energy in a shorter period of time, it is possible to get around the Solar System rather quickly. *Distance* in the Solar System means very little; the major factor is *velocity* and, for manned vehicles, the *time* required.

It is possible to get nearly anywhere in the Inner Solar System (from the Planet Mercury out to the Planet Jupiter, for example) if the spaceship has the ability to change its velocity by only 15,000 feet per second. This is no big deal; it is about half of the velocity required to go into orbit around the Earth, starting from the Earth's surface. The various changes in velocity required to get from place to place in the Solar System are shown in the accompanying table.

In 1963 Dandridge M. Cole, one of the pioneer futurists who was working for General Electric at the time, laid out the basic requirements for a "Space DC-3." This is now called "space tug," and it is much smaller than Cole's proposal. The

Space DC-3 has characteristics that are detailed in the accompanying table. It would have the capability of transporting up to 100 tons of payload anywhere in the Solar System out to the orbit of Jupiter. A Space DC-3 could be manned or unmanned. Cost for shuttling payload around in increments of 15,000 feet per second amounts to about 17 cents per pound for rocket propellants. Nothing new in the way of vehicle technology is required for this. The propellants are liquid hydrogen and liquid oxygen. Both are quite likely to be available in quantity in space itself through the application of solar energy.

However, a cargo need not be enclosed in a vessel in order to be transported around in space. The cargo can be the vessel itself because it really needs no container. The "motor" of a Space DC-3 could well be a small, manned propulsion unit with a small cabin and lot of propellant tankage. It's the space-going version of a locomotive. For a payload going from the planetoid belt to the orbit of the Earth-moon system, the space locomotive would accelerate the payload to the proper velocity—perhaps taking several days to do it. Once the payload was traveling at the right velocity in the right direction, the locomotive uncouples and comes back; it will take considerably less rocket propellant to do this because the locomotive is small and light compared to the payload it had been accelerating. Meantime, the payload continues on its flight.

The flight could be very long indeed. Economics may dictate that the payload is put into a highly economical, low-energy "Hohmann transfer trajectory." This is a type of interplanetary flight path that requires the least amount of rocket power, and therefore the least amount of expensive rocket propellant, to get from planet to planet. However, time is not of the essence, and the Hohmann transfer trajectory between planets is about as long as a flight between two worlds can take. It might take several hundred days to make the journey from the planetoid belt to Earth. Who cares? If time is not of the essence, the only thing that will be of interest to the space-going purchasing agent and inventory con-

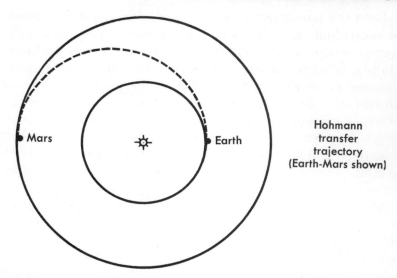

Hohmann transfer trajectory (Earth-Mars shown)

trol person is the schedule of arrivals of the payloads at the other end of the flight. With such an astronautic "pipeline" in operation, it makes little difference how long a load of planetary iron ore is in transit; all that matters is the constant delivery rate.

When the load arrives near its destination, it is met by another space locomotive that couples to it and changes its incoming velocity by application of rocket power in just the reverse manner used in getting it started in the first place.

Studies have already shown that such a "space locomotive" approach saves time, energy, and therefore money. A manned space freighter must take its crew along with it as it accompanies or encloses the payload during the long trip; the crew uses food, oxygen, and other consumables that relate to energy consumption. A crew is required for every load. However, using the space locomotive concept, fewer crews are needed and their requirements for breathables, food, and other life-support equipment are less; they are in the locomotive for a shorter period of time and can replenish their life-support consumables more often from the central space terminal from which they operate.

Over the past quarter of a century, a tremendous amount of individual work and study has gone into the subject of space transportation. Books have been written about it. Highly technical papers have been presented at meetings of the American Rocket Society, the American Astronautical Society, the British Interplanetary Society, and dozens of other organizations around the world. Technical journals devoted to space travel are treasure troves for space artists who must come up with new spaceship designs to illustrate stories. There have been hobby model kits of workable space vehicles of the future. In short, spaceflight has stimulated the conceptualization of vehicles to a far greater extent than any previous mode of transportation. But there have been very few attempts to bring the whole thing together for consideration as a system. In nearly every case, space transportation has been considered as a means for getting *explorers*—scientists and technical people—around the Solar System for informational purposes. The studies are equally applicable to the problems of space transportation in support of a space industrial complex, and more valid from the point of view of end-use utility.

The big bugaboo of the space industry concept, space transportation, really isn't that much of a cost problem after all. A decent, manned Earth-to-orbit space transportation system will cost about the same to develop as a modern coal-fired electrical generating plant—the big Navaho generating station in the Four Corners area of the American West has a price tag of $630 million, for example. The vehicles themselves can cost less than a Boeing 747. They can place a payload into near-Earth orbit for as little as $10 per pound.

Getting around in the space environment once we are out of the Earth's gravity field costs even less.

And we don't have to invent anything new to do any of this.

We just have to do it.

And the Third Industrial Revolution provides the incentive.

Space Transportation Velocity Requirements
(From D. M. Cole, "The Next Fifty Years in Space, the
Transportation Problem")

Low Earth orbit to escape:	11,000 ft/sec
Low Earth orbit to high lunar orbit:	12,000 ft/sec
Low Earth orbit to Mars orbit:	13,000 ft/sec
High lunar orbit to lunar surface:	8,000 ft/sec
High lunar orbit to Mars orbit:	8,000 ft/sec
Surface of Moon to Mars orbit:	15,000 ft/sec
Surface of Mars to low Mars orbit:	15,000 ft/sec
Low Mars orbit to low Earth orbit:	10,000 ft/sec
Surface of Mercury to low Mercury orbit:	11,000 ft/sec
Low Mercury orbit to low Venus orbit:	12,000 ft/sec
Low Earth orbit to low Venus orbit:	12,000 ft/sec
Orbit of Venus to orbit of Mercury:	9,000 ft/sec
Low Mars orbit to planetoids:	15,000 ft/sec
Planetoids to orbit of Jupiter:	10,000 ft/sec

Solar System Deep Space Ship
"SPACE DC-3"
(From D. M. Cole, *op. cit.*)

Gross mass:	1,000,000 lb
Propellant mass:	675,000 lb
Payload mass:	250,000 lb
Structure mass:	75,000 lb
Propellant fraction:	675/750=0.90
Propellants:	Liquid hydrogen, liquid oxygen
Specific impulse:	420 lb-sec/lb
Exhaust velocity:	13,500 ft/sec
Total mission velocity:	15,000 ft/sec
Propellant cost per pound of payload (at 5 cents per pound for LH and LOX)	17 cents

CHAPTER

4

The Black Beyond

In 1803 Thomas Jefferson forecast that the newly acquired Louisiana Purchase might be fully settled by the year A.D. 2600. And much later, even after Lewis and Clark had made the initial explorations and settlers were poling up the Missouri River to new lands, a U.S. Senator from Indiana felt it would do no harm to provide the fledgling Western railroads with land grants. "These lands are classed as refuse," he stated.

On December 25, 1968, *Apollo-8* Astronaut Frank Borman looked down on the Earth's Moon and remarked, "It's not a very good place to live or work." Standing on that same Moon in July, 1969, Edwin Aldrin remarked, "Magnificent desolation." As desolate and seemingly useless as the endless flatlands of the Louisiana Purchase must have seemed to Lewis and Clark. But not useless—neither the Great Plains nor the Moon.

Lewis and Clark, Zebulon Pike, John Charles Fremont, and the others knew not where their journey would take them. We do.

The Solar System is Earth's own backyard. For centuries, people have watched the planets move across the sky. They have studied them through telescopes. They have only recently sent robot explorers to probe the mysteries of the planets from close at hand. Where machines can go and man can also go, man will go.

The Solar System contains one average-sized, middle-aged

star, nine major planets, thirty-four major planetary satellites, four major satellite-sized planetoids, an estimated 40,000 planetoids, and an unknown number of comets and other bits of sky junk. Current theory leads us to believe that all of this was formed out of the same proto-material approximately four billion years ago, give or take a few hundred million years or so.

Most of this material is located orbiting in a flat plane around the major star called the Sun. To our earthbound minds, the distances between objects in the Solar System are so large as to be meaningless when couched in normal terms such as miles or kilometers—earthly measurements intended for use by people walking around on a planetary surface.

Because distances are so great in the Solar System, we should use a measure of *velocity* and *time* rather than distance. We are already starting to do this on Earth's surface, by the way. Distance on Earth is no longer as important as the *time* it takes to cover the distance. Ask someone, "How far is it to town?" The answer is likely to come back, "Oh, about thirty minutes or so." Transcontinental distances are now "four hours." And it is "six hours" to Europe. These time/distance statements are based upon the usual speed of the transportation device used to make the journey—a car or a jet plane, in our examples.

The size of the Solar System can be shrunk to terms that can be more readily grasped if we measure it in terms of the time required for light, traveling at 186,000 miles per second, to make a journey among the planets. Light is an electromagnetic radiation just like heat, radar, and radio. We are already using the speed of light in our early exploration of the Solar System because we are communicating with spacecraft by radio, measuring their distances with radar, and controlling them by radio signals as well.

Traveling at a speed of 186,000 miles per second or about 300,000 kilometers per second, light from the Sun takes about 2 hours to reach the planet Pluto on the outermost marches of the Solar System. Therefore, Pluto's distance could be stated as "two light-hours."

Earth is located about 8 light-minutes from the Sun, while

the Moon is about 2 light-seconds from Earth. We had very little trouble handling the 2 seconds time lag in Earth-Moon communications with Apollo astronauts, and the lag was hardly apparent to TV watchers. It's a different story for Mars which, at its closest approach to Earth, would require about a 6-minute delay between sending a message and getting a reply; telephone conversations between space vehicles around Mars and home base on Earth are going to be too lengthy for convenience; telex or TWX would seem to be the better bet.

How much room is out there? Depends on how you look at it. In terms of land area, however bleak and forbidding it looks today (remember the Louisiana Purchase), there are three other planets or major satellites to count on—Mercury, Mars, and the Moon. Venus appears to be too hot to be of immediate use with technology that we have today or can reasonably forecast in the next one hundred years.

Let's compare these to the Earth in terms of size and usable surface area.

Earth is roughly 7,900 miles in diameter and has a land area of about 57 million square miles. Over 70 percent of the Earth's surface is covered with water and is not counted here.

Little Mercury, closest to the Sun, is about 3,000 miles in diameter and sports a surface area of about 28 million square miles.

Ruddy Mars, the first planet we are likely to visit, is 4,220 miles in diameter and has a surface area of about 56 million square miles.

Our own Moon is 2,160 miles in diameter with a surface area of 14.6 million square miles.

Sitting out there on only *three* of the bodies of the Solar System is a surface area 1.72 times that of Planet Earth! Mars alone has nearly the land surface area of Earth.

We are not talking about the Solar System being filled with a planet called Earth and some other small and insignificant rocks. There are whole planets out there, some with surface areas almost as great as Earth's.

A real estate bonanza, but can we live there?

Yes—if we can keep people alive for hours on the surface

of the Moon, we can keep them alive anywhere in the Solar System. If we can keep people in a can called *Skylab* for 84 days at a time, working, living, eating, and happy, we can keep them alive, working, living, eating, and happy anywhere in the Solar System.

Industry today keeps people alive, working, and happy miles underground, in the heat of the African desert, in the bitter cold of the Alaskan arctic, and in other terrestrial environments just as hostile, just as forbidding, and just as difficult as the surface of the Moon.

How about raw materials for industry? In the beginning, with Space Shuttle and the early in-orbit space industries of the next decade, most of the raw materials required will be lifted into space from the Earth's surface. But, even at costs of $10 per pound to lift mass into orbit, it is quickly going to get far too expensive to boost raw materials for space industry from Earth's dwindling supplies into orbit. One of the reasons why space industry will get off the ground, so to speak, is the fact that Earth's supply of raw materials is finite and dwindling rapidly. Space industry will not further deplete them for very long because of this and because of the high cost of transporting raw materials into space from Earth.

There will soon come a time when space industry requires raw materials in quantities carried by Great Lakes ore carriers that transport up to 20,000 tons at a time. That's 40 million pounds, and at $10 per pound to lift it from Earth, you get it somewhere else that's cheaper.

Like the Moon. Or the planetoid belt between Mars and Jupiter.

Will we be able to find the raw materials we need for a space-based industrial complex on the planets of the Solar System? On the Moon? Perhaps. We don't know yet. We have not explored them thoroughly.

We do not yet know that the Moon is only a worthless piece of selenology. We have landed only 12 men on its surface at 6 points. Only one of these men was a trained geologist. The surface of the Moon is about one-fourth the surface of the Earth. Would it be possible to make a geological survey of

Planet Earth by landing at 24 points with 48 men, only 4 of whom were geologists?

It takes years of crawling across the landscape on foot and in wheeled vehicles to complete a thorough geological survey of a single Western state.

We cannot yet rule out the Moon as a source of raw materials for space industry. And if they are there, they can be hurled into space at a speed of a little over a mile per second; this will put them in space forever, where they can be used by industry in space.

Getting a payload off the surface of the Moon and into space is not the expensive, straining, thundering job that it is on Earth. The Moon does not have an atmosphere. And the Moon has only one-sixth the surface gravity of Earth.

Engineering designs already exist for a lunar surface catapult capable of completely eliminating the use of rocket power for hurling large payloads into space from the Moon. This catapult would be built right along the lunar surface. Since there is no air on the Moon, there is no need to launch a payload vertically to get through air resistance as rapidly as possible. A mountain catapult would be required on Earth to lob a space vehicle up through the atmosphere on as short a path as possible to space. The lunar catapult need provide a boost of only about one and a half miles per second (2.4 kilometers per second); this is the speed of small, single-staged research sounding rockets commonly launched on Earth. Power will come from a series of solar power screens covered with silicon solar cells. The catapult will be a "linear electric motor," an electromagnetic accelerator.

How about raw materials on the other planets? It may be far too costly and far beyond our present technology to get any sort of raw materials from the Planet Venus, not only because of the high temperature, but also because of the very thick atmosphere and the high surface gravity (almost equal to that of Earth).

Mercury? Perhaps. We have only flown past it with unmanned spacecraft carrying cameras. We have yet to explore it.

Mars? Perhaps. Same reasons. But a very good chance that

the red color of parts of the planet may be iron oxide, which means a source of iron, the main material of our technology. We will know more by the turn of the century after men have looked it over on the spot.

The major planets—Jupiter, Saturn, Uranus, Neptune? We have yet to do more than send an unmanned spacecraft flying past Jupiter. The outer planets are "gas giants." They may be made up of hydrogen, methane, ammonia, and other hydrocarbons. If so, they could be the "mother lodes" of raw material for space-based chemical industries. After all, regardless of how well we manage Earth's store of hydrocarbons—petroleum and coal, primarily—sooner or later they will be exhausted. The gas-giant major planets will probably be the chemical sources of future industry. And there is a lot of it out there, especially in Jupiter with a diameter of about 88,800 miles. That's a volume of 2,933,100,000,000,000 cubic miles—a lot of hydrogen, methane, ammonia, etc.

But Jupiter is a large, massive planet and has an escape velocity almost 5.5 times that of Earth. If raw materials are going to be so expensive to lift out of Earth's relatively shallow gravity well, isn't Jupiter's far too strong and, therefore, far too expensive? Perhaps. But in the case of the Earth, the raw materials come from its solid surface and have to be lifted from zero velocity. It is Jupiter's *atmosphere* that will be "mined," not its somewhat dubious surface. It would be entirely possible to design atmospheric mining vehicles that would orbit around Jupiter and, with a slight change of orbit, swoop down to graze the tops of Jupiter's atmosphere, filling their chemical payload tanks through nose scoops that would make a guppy ashamed. Such ships would then swoop back out of the jovian atmosphere and, with a small "delta-vee" application of rocket power, take up orbits around Jupiter again.

Although it is possible with current and conceivable technology to seriously consider "mining" the chemical atmospheres of Jupiter and Saturn—and also Uranus and Neptune, when we find out more about them—the jovian and saturnian satellite systems may also provide a bonanza of raw materials.

Jupiter is accompanied by twelve (at last count) satellites and Saturn herds ten. They are big, as moons go, some of them being as large as or larger than some of the inner planets. Jupiter's Ganymede is over 5,000 miles in diameter, larger than Earth's Moon and larger than the Planet Mars. We know very little about these massive moons at this time, but we will find out more about them in the years to come. They may be rocky bodies; if so, they may still be valuable raw material sources. However, some astronomers have advanced the theory that the jovian satellites may have small rocky cores surrounded by deep layers of liquid—ammonia, methane, etc. If so, these satellites will be very valuable indeed.

But by far the richest Golconda of raw materials in the Solar System is likely to be the planetoid belt (often called the asteroid belt, although it is not composed of "small stars" as that name would have you believe).

The planetoid belt lies between the orbits of Mars and Jupiter. It contains between 30,000 and 40,000 small bodies ranging in size from the planetoid Ceres (477 miles in diameter) down to sky junk the size of boulders and large rocks.

Most of these planetoids are so small that their escape velocities can be measured in feet per minute rather than miles per second. It would be possible to leap off some of them and soar into space forever. They don't have real gravity "wells" as we think of them for planets; they are more like gravity "potholes" in the road of space. It will require very little in terms of velocity change to land on one and take off again. As a matter of fact, the problem is likely to be one of managing to stay on the surface with such low gravity!

What are the planetoids? Why isn't there a major planet there instead of a broad belt of sky junk? We don't know, and it really doesn't make too much difference from the point of view of the Third Industrial Revolution whether or not we find out. The important thing is that they are there. They may be part of a planet that never formed four billion years ago because of the gravitational perturbations of giant Jupiter, which kept the sky junk from getting together as it did for the other planets. Perhaps it was once the Planet Lucifer that formed and then came apart for unknown reasons, perhaps under the gravitational pull of Jupiter.

The satellites of Mars are about the same size as the average planetoid; they may be planetoids that were captured by Mars. The photos taken of them by the *Mariner* spacecraft showed them to be rocky, irregular, crater-pocked objects. Most of the planetoids will probably look the same.

Not all of the larger planetoids are spherical or near-spherical. The planetoid Eros is apparently shaped like a rough sausage about the length and diameter of Manhattan Island. Some ten other planetoids exhibit marked changes in the brightness of the sunlight reflected from them, and astronomers theorize that these planetoids are also large, irregular bodies like Eros.

This has led astronomers to further theorize that these big planetoid rocks have chemical compositons much like the other rocky bodies of the Solar System. We can at this time only theorize regarding what the planetoids are made of. However, chances are very good that their chemical makeup will resemble that of the meteorites that strike the Earth.

Meteorites fall into two general types: the rocky or chondrite meteorites that make up the majority of the meteors that reach the surface of the Earth and the metallic meteorites that occur in fewer numbers.

The meteor that blasted the Barringer Meteor Crater out of the Arizona landscape was a metallic meteor about 80 feet in diameter.

The composition of the chondrite meteorites seems to be reasonably uniform from one sky rock to the next. On the other hand, nickel-iron meteorites are just that: nickel and iron of very high quality.

If the planetoids are made of the same stuff as the chondrite meteorites, their general, normal chemical composition would be as shown in the accompanying table. They would consist mostly of silicates, but over 13 percent of their mass would be metallic—iron, aluminum, magnesium, manganese, and titanium.

The average-sized planetoid may be about one mile in

diameter. In this hypothetical "standard" one-mile-diameter planetoid with a chemical composition identical to that of the chondrite meteorites, there would be 22 million tons of iron.

This is *seven times* the average mid-1974 weekly steel output of the United States. And our hypothetical planetoid is only one of thousands of planetoids.

If, on the other hand, our one-mile-diameter planetoid had the same composition as a nickel-iron meteorite, it would contain 33 billion tons of nickel and iron. That's a *230-year* United States production supply at mid-1974 levels. This astronomical Mesabi Range is orbiting the Sun between Mars and Jupiter. It is made up of chunks ranging in diameter from several miles down to boulder size.

But what can we do with it in its location between Mars and Jupiter?

Probable Composition of the Planetoids
based upon the composition of chondrite meteorites
(From D. M. Cole, "Applications of Planetary Resources")

Component	Percent
SiO_2	38.29
MgO	23.93
FeO	11.95
Al_2O_3	2.72
CaO	1.90
Na_2O	0.90
K_2O	0.10
Cr_2O_3	0.37
MnO	0.26
TiO_2	0.11
P_2O_5	0.20
H_2O	0.27
FeS	5.89
Total silicates	81.00
Total metals	13.11

We can move it to another location in the Solar System if we want to in order to process it closer to its point of use, if needed. We have the rocket technology today that would allow us to move a one-mile-diameter planetoid from its orbit between Mars and Jupiter to an orbit around the Earth. Nudge it with rocket power to get it moving in toward Earth, then nudge it again in orbit around Earth to keep it there.

We may not have to move a planetoid to get the iron and other metals out of it. Using concentrated and focused solar energy, it would be possible to smelt the iron out of the planetoid in the planetoid belt itself. But it may be cheaper and quicker to do it closer to the Sun—say, in Earth orbit—because of the higher density of solar energy closer to the Sun.

Iron is the basic foundation of our technical civilization. If we have iron and energy, we can do nearly anything. And there appears to be iron in space, waiting for us to come and get it and use it.

What else is out there in the way of raw materials?

We have just started to look. We have landed men on only one other celestial body. We have sent remote-controlled spacecraft to four others and gotten some data back, but from distances of thousands of miles. We will have to have a closer look to find out exactly what is there.

If our only knowledge of Earth's natural resources were based only on a few scientific observations from orbit and from two or three quick flybys, it is extremely doubtful if we would have a complete catalog of these resources. We must get down on the ground, pick up rocks, look into nooks and crannies of the landscape, and dig in order to do real prospecting. Therefore, we are going to have to have much closer looks at the planets and planetoids to determine their values as sources of raw materials.

But does anyone wish to bet that we will *not* find gold, silver, zinc, copper, lead, tin, molybdenum, and other industrial metals there? Does anyone wish to bet that Earth is the *only* planet in the Solar System that contains these metals?

Lewis and Clark walked right over the mineral wealth of Montana and didn't even suspect that it was there. I seriously doubt that we will make the same mistake in the Solar System.

The planetoid belt is awaiting a future Andrew Carnegie.

CHAPTER

5

Energy Enough for Everything

"MANKIND HAS BEEN converting natural energy into social organization; as he has drawn more and more energy from Earth's natural storehouse, he has organized himself into social institutions of increasing size and complexity."

If this energy thesis of anthropologist Carleton Coon is correct, and if we do indeed face an energy shortage here on Earth because of depletion of natural fossil fuel resources, we must either develop—fast—nonfossil fuel sources or go looking elsewhere for more energy.

We will probably do both.

Considerable effort is now being expended to develop the hydrogen fusion process as a controlled energy source. We've already done the trick with the thermonuclear bomb, the so-called H-bomb. But that is rather sudden, and adapting it for a steady energy source is somewhat like trying to harness the energy of a lightning bolt.

Happily, we have a friendly neighborhood nuclear fusion reactor operating very well. It's the Sun, and the neighborhood is the Solar System. There is much talk today about solar energy, but it will really come into its own in space.

The Sun never sets out there.

Is it enough? Is it the only energy source useful in space? After all, if we can get into space economically and if we ca find raw materials to use out there, is there enough do anything?

64

Solar energy is only one of several possible energy sources in space, but it may well be the most important because it is there and it is available. Lots of it.

The Sun is approximately 865,000 miles in diameter. It is an average-sized star. We are learning more about it all the time. It is a natural hydrogen fusion reactor that works. It has a surface area of 23,177,000,000,000,000,000,000 square yards. Each square yard radiates energy at the rate of about 80,000 horsepower *continuously.*

At a distance of 93 million miles from the Sun, this radiant energy produces roughly three-tenths of a calorie per square inch per minute. This is the same calorie watched by weight watchers, the amount of energy required to heat one gram of water one degree Centigrade (or about one-thirtieth of an ounce of water about two degrees Fahrenheit). This is the amount of solar energy that is available at the distance of the Earth's orbit around the Sun, the amount that falls on every square inch of the Earth's surface every minute. It doesn't sound like very much, does it?

But you can really concentrate this if you can build a mirror big enough.

They've done so in France and in several places in the United States. This is the way they generate extremely high temperatures for scientific research.

As a matter of fact, you can use this three-tenths of a calorie per square inch per minute to start a fire in a hurry using a small magnifying glass.

In space, where everything is in a weightless condition, it is possible to build a very big mirror very easily with very simple materials.

Take a load of thin aluminized Mylar sheet into space. Build a very simple and very spindly frame. Attach the aluminized Mylar to the frame. Start it spinning like a Frisbee disc. Spin it into a parabolic shape. You can make it very large indeed.

If it is about 300 feet in diameter, it will concentrate 11 megawatts of solar energy at its focus.

If this energy is concentrated in a target about 4 inches square, it produces enough heat to *vaporize* over 24 pounds of copper *per second*.

This is obviously energy availability of industrial magnitude. And it is there now, waiting to be used. And it will be there for millions of years. There is no shortage of solar energy in space.

If your industrial process or space factory needs more energy, locate it closer to the Sun. It's a lot hotter there.

As your space factory locations get farther from the sun, they will require larger solar mirror collectors. More Mylar. A bigger frame.

There will probably come a time at some far future date—say, in the twenty-fifth century, perhaps—when the natural energy needs of the human race—growing as they will continue to in order to sustain larger and more complex social institutions—may require the total energy output of the entire Sun. We can't do that yet with any technology that we can forecast or envision today, but it may result in a man-made "black hole." But when we get to the point technically where we can create our own black holes in space, we will also have the ability to use the energy output of other stars, too. It's not something we will have to worry about for several centuries yet; there is plenty of solar energy to meet our needs for anything that we can accomplish in the next century.

We will, of course, create our own nuclear energy sources to take into space to provide energy there. We can do this today with nuclear fission reactors if we wish, and we have already emplaced several nuclear electric generators on the surface of the Moon in Project Apollo.

The unfortunate thing about today's nuclear fission reactors is the high level of ionizing radiation they produce while in operation. On the Earth, we can provide shielding against this radiation, and the only shielding that is effective is mass. Lead. Weight.

Weight (mass) is and will continue to be a problem in space travel. It takes propulsive energy in the form of rocket propellants to change the velocity of mass. The greater the mass, the more energy required.

For unmanned space factories, nuclear reactors may well be provided as energy sources.

But these reactors will have to be shielded if people are to visit or remain in these space factories.

Shielding is weight.

Some shielding will be required even for unmanned applications because of the fact that radiation often has a profound effect on other materials.

However, engineers are trained to take disadvantages and turn them into advantages if they can.

Nuclear reactors in space may be used not only as energy sources, but also as industrial tools because of the effects their ionizing radiations have on materials, on chemical processes, and on biological processes. More of this later.

The third energy source that may be useful for space industry is chemical energy.

This may appear at first glance to be an incorrect statement. Most of the chemical energy sources with which we are familiar are the oxidizer-fuel exothermic reactions that proceed by combustion. On Earth this requires a fuel to burn and atmospheric oxygen to sustain the reaction; heat is produced. It is a common energy source on Earth because of the very large supply of atmospheric oxygen. In space there is no atmospheric oxygen; there is no atmosphere.

But we can still use chemical energy sources, and there are several of these that are likely candidates.

The combustion of hydrogen and oxygen is one of the most energetic reactions available. The reaction produces water and heat. Hydrogen is the fuel, and oxygen is, naturally, the oxidizer.

But how do we get hydrogen and oxygen in space?

From water.

There are two possibilities. Both start with solar energy collected and focused by a large mirror.

The long, hard way around the car barns to get hydrogen

and oxygen may be termed the "steam cycle." Solar energy is used to heat water and produce steam. This steam then drives a turbogenerator to produce electricity. Some of the electricity is used to pump the water through the solar boiler in the primary steam loop. The rest of the electricity is used to dissociate water by electrolysis. This is one of the simplest of freshman chemistry experiments. The passage of electricity through water causes hydrogen to appear at one electrode and oxygen at the other. The water molecule has been broken down by electrical energy. The hydrogen and oxygen are then stored. To get the energy out of this system, one must bring the hydrogen and oxygen back together again in a "reactor" of sorts where combustion takes place. This produces heat energy and water. The water can be put back through the system and broken down again.

The problem with this complex cycle are its losses because solar energy is used in a very inefficient way.

A possible shortcut around the use of a steam turbogenerator would be the use of solar cells to produce electricity directly, and the electricity is then used in the electrolysis process. This is even more inefficient because of the solar cells. They aren't very efficient converters of sunlight to electricity. Much work remains to be done with solar cells to increase their efficiency and to improve their working life. They should not be ruled out.

The most direct way to convert solar energy into chemical energy in space would be to use the solar energy to heat the water to about 3000° F. At this high temperature, the water molecule dissociates into hydrogen and oxygen because of the very high temperature. The high-temperature dissociation route may eventually be the one most frequently used in space for chemical energy purposes, but it is the one least understood at this time and the one on which most development work will have to be done.

Burning hydrogen in oxygen to produce heat and water is only one way to utilize a chemical reaction, if you want heat. If you want electricity out of the process, the way to go is via the fuel cell.

Fuel cells have a long history of providing electric power in space. They were used on the *Gemini* space capsules and on the *Apollo* vehicles. Through use of clever design, hydrogen and oxygen are brought together in a fuel cell to produce electricity directly with water as the end product. An H-O fuel cell is fairly efficient. We have seen them used in space already, and we will continue to see them used in space in the future.

All of these chemical energy systems start with solar energy. And we may see some interesting inverse technical feedback here. There has been much talk about developing solar energy on the Earth's surface to ease the growing shortage of natural energy sources. The big problem with solar energy is storage; at night or when it is cloudy, one cannot use solar energy. What we need is a "bucket" to store solar energy in until we want to use it. Dissociating water into its hydrogen and oxygen components provides a way to do this. When you put them back together to form water again, the energy required to break the water molecule apart in the first place is liberated. Hence, we have a "bucket" to carry solar energy around in and to store it in until nighttime when we would want to use it to make electricity to light our homes and heat our buildings. Earthbound developments in solar energy over the next several decades may well provide us with ready-made solar energy techniques for use in space later in this century!

Or if it happens to work the other way, the development of solar energy systems for space industry may solve the solar energy problem here at home on Earth.

It will be interesting to see which one happens.

There are many other sources of both electricity and heat energy in chemical reactions that do not involve the oxidizer-fuel combustion process. Space engineers will eventually want to take a look at all of these. For example, if you take a bit of zinc chloride, put it in a bottle, and add water to it—it makes an excellent flux for soldering—the resultant liberation of the heat of solution can get the bottle so hot it will burn your hand. This is just an interesting little parlor scientific trick right now; tomorrow something like it could be providing heat energy for space industry.

As an interesting sidelight on this discussion of energy for space industry, remember that we earlier discussed the causes for the First Industrial Revolution. It is possible to consider that the First Industrial Revolution was triggered by an energy shortage: the final cutting of England's hardwood forests and the resulting severe shortage of fuel. The development that arose as a result of this required the extensive use of fossil fuels that had been stored up in the Earth for eons; a forest can be regrown in less than half a century, but the fossil fuel reserves of Planet Earth could not be replaced within a reasonable period of time—less than a couple of hundred million years, at least. Fossil fuels such as coal and petroleum are limited because the size of the Earth is finite; there's only so much of them underground for us to dig up. We're running out, whether it finally comes to an end in a dozen years, as some pessimists claim, or hundreds of years, as the optimists state.

Is there an analogy between the depletion of a fuel source at the start of both the First and the Third Industrial Revolutions? Will the fossil fuel crisis result in the Third Industrial Revolution? Perhaps, and it might be very easy to hang on this the full justification for the Third Industrial Revolution. But it's only one of the factors.

Cause or solution, the development of energy systems for space industry may solve the fossil fuel energy crisis here on Earth. There is plenty of energy out there, waiting to be used. A whole star, right in our own backyard.

As a coda to the energy discussion, what do you do with this energy when you're through with it? Any engineer will explain a principle called the Second Law of Thermodynamics. In simple terms, this principle states that you can't achieve 100 percent efficiency; there will *always* be some waste energy that you cannot possibly use. You can see an example of this at any petroleum refinery where stack gas is burned off in a bright, wavering orange flame; the heat energy of this stack gas is so low and its pressure is also so very low that it would take more energy than the stack gas possesses to harness it. It is the unusable residue, the cinders and ash of the petrochemical operation. The Second Law of Thermodynamics is not a law that can be repealed by Congress; it is an inevitable consequence of the universe.

Since every industrial operation therefore has waste energy left over that cannot possibly be used, what can you do with it? Get rid of it. That's the only thing that can be done.

On Earth it can be discharged into the atmosphere or into the nearby river. These are called "heat sinks." Fortunately, both the atmosphere and most of the bodies of water that were handy have been able to absorb this waste energy without upsetting things—until recently when industrial operations started to outgrow the ability of the environment to handle the waste. We've started to run out of heat sinks, which is another reason to start thinking about moving industrial operations out into space.

What can we do with the waste heat out there? Can we pollute the Solar System with our waste heat?

Although some space industrial operations may be located on other planetary surfaces and may therefore be able to use some of this planetary mass as a heat sink, there are only two basic methods of discharging heat from a body in space or on the surface of an essentially airless celestial body. These are (1) by radiation and (2) by dumping overboard into space an expendable coolant mass that contains the waste heat.

The latter method is wasteful; we could use it in the early manned satellite programs. But when it comes to keeping vehicles in space for extended periods of time, we run up against a limitation on the mass of coolant that we can carry along.

To get rid of waste heat in space, we must radiate it back into space itself.

Space heat radiators will be very large assemblies with a great deal of surface area—acres for the small ones, square miles for larger ones. To keep them from collecting solar energy on the sunward side, they will be mirror-surfaced on that side. The other side will point away from the sun and toward interstellar space; it will be rough and dark. Waste heat from the industrial operation will be carried to the radiator by means of heat pipes or by liquids such as molten sodium or mercury. To operate most efficiently, the radiator will be operated at the maximum temperature permitted by the materials of which it is made. As the technology of high-temperature materials progresses, this radiator temperature will be increased. Some of them will glow a dull red. They will be discharging waste heat into space.

Because of the waste heat problem and the need to get rid of it into space, every calorie coming into a space installation from the sun will have to be carefully and completely accounted for. If this is not done, the space complex is going to get very warm, very quickly. It could end up being totally unlivable and completely nonfunctional in a surprisingly short period of time. It is quite unlikely that it would melt down to a white-hot glob; most of the heat-transfer, collecting, and processing equipment would be completely out of operation long before that point was reached.

How about using insulation as we do here on Earth? First of all, insulation must be looked upon as a time delay for heat. It doesn't stop heat from getting through it; it slows down the rate at which heat penetrates. A well-insulated home on Earth will get extremely warm in the desert sun after a week or less, and it will also get very cold in the wintertime after several days have gone by. Insulation delays the transmission of heat. It will be used in space as a heat delay material.

On the other end of the scale, we have the modern, Space Age heat pipe, which is a super heat conductor. Basically a very simple gadget in appearance, a heat pipe is a very sophisticated device. It is a closed, evacuated tube lined with a capillary wicking material that is saturated with a volatile liquid. When one end of the heat pipe is warmed, this volatile liquid flashes into vapor that travels down the hollow middle of the tube to the cool end where it condenses, releasing its heat. Today's heat pipe can transmit more than five hundred times more heat than a solid metal rod of the same cross sec-tion. Heat pipes will certainly be used in space complexes to transmit heat rapidly and efficiently.

It is quite possible to fully and completely design space complexes to handle both the heat input from solar energy collectors and the waste heat that is an inevitable consequence of the Second Law of Thermodynamics. It's just engineering—which is fiddling with finicky figures for months on end to make sure everything is going to work—and then working in a "Finagle Factor" (sometimes called a "safety factor") of as much as ten to make sure that (1) you haven't goofed somewhere, (2) that somebody isn't going to goof putting it together, (3) that something isn't going to quit and create a disaster, or (4) that some dum-dum isn't going to goof and precipitate a complete collapse of the system. For this reason, heat radiators are going to be the most prominent features of space complexes and space factories. They will be huge because we no longer have to worry about having them punched full of holes like a colander by meteors. The meteor hazard in space has been shown to be practically nil. (There will be repairs required to patch the little pinholes that micrometeors will punch through the heat radiators, however.)

Will this pollute the Solar System with excess heat? The answer here is a flat, unequivocal NO.

The heat originally comes from the Sun. The space complex merely interrupts a minuscule fraction of the total heat energy the Sun continuously radiates into the Solar System and eventually into interstellar space. Our space factories will not be adding any heat to the environment out there.

It doesn't look as if there is any way we can heat-pollute the Moon from any industrial operations there, either. Again, waste heat will have to be radiated away into space because the Moon has no sensible atmosphere to pollute.

And there is no way known to anyone right now that would dump more heat into the Earth's atmosphere from a space factory.

Space industry is going to be based almost totally on some form of solar energy. It will only interrupt on a minute scale the enormous energy output of the Sun. That energy that cannot be used will go back into the universe where it started in the first place.

The energy situation in space—its availability and the existence of more-than-adequate heat sinks—is another reason why the Solar System is the best possible place for industrial operations, certainly much more compatible with the surroundings than it is on Earth.

CHAPTER

6

Room with a View

ON EARTH, industrial plants, operations, and complexes are usually located close to some critical requirement for the particular operation, and the environment in which the industrial process takes place is then created in the facility that is built. For example, steel mills are always located near a source of coal, and the iron ore is transported to the coal because more coal than iron ore is used. The blast furnace is an example of a created environment for an industrial process. On the other hand, copper smelters are usually always located at or near the copper ore supply because natural gas and other fossil fuels can be easily transported cheaply to the smelter beside the big open pit, whereas it takes lots of ore for a very little amount of copper.

The situation is likely to be a little bit different for space industry. The location of the industrial operation will probably depend upon the environment needed for the industrial process itself. One reason for this is the fact that solar energy is reasonably easy to come by anywhere on the sunward side of the planetoid belt; except for very special processes requiring a great deal of energy, an industrial operation in the inner Solar System can be located near the raw material source or in a part of the Solar System where it will be easiest to handle the particular space environment needed for the process.

Unlike on Earth, where artificial industrial process environments must be created, in space the industrial facility is built around the industrial environment.

What do we mean by this? Simply that the open spaces of the Solar System offer us the widest variety of industrial environments we have ever had the opportunity to utilize.

This is based on the earlier statement that outer space must be looked upon as a useful place instead of a place in which to spend money searching for scientific truths.

When we take a close look at the known physical environment characteristics of the open space of the Solar System with this in mind, we discover that the following characteristics are of interest to industrial engineers:

1. Gravity-free environment and its corollary, the controlled acceleration environment, which includes the controlled gravity gradient environment as well.

2. Matter-free (high vacuum) environment and its antithesis, the controlled matter (controlled pressure) environment.

3. The controlled radiation environment with the ready availability of wide-spectrum electromagnetic radiation of a wide variety of intensities.

4. The wide-range temperature environment with readily available temperatures ranging from nearly absolute zero up to nearly solar surface temperatures.

5. Wide-range energy density environment wherein we can readily control with relative ease the amount of energy going into a process in a given period of time.

These five environmental characteristics can be obtained on the Earth's surface, but only for brief periods of time or at great cost. (Often, this cost is in terms of human peace of mind; what person working around high-vacuum industrial processes here on Earth has ever gotten used to the maddening *plop-plop-plop* of vacuum pumps all day long?) However, this ability to create the conditions here means that we do not really have to wait until we have a full-fledged orbital industrial research laboratory staffed and ready to go in order to begin investigating what we might be able to do in the space environment.

For many years, the people at NASA's George C. Marshall Space Flight Center in Huntsville, Alabama, have been studying space manufacturing processes; there have been a number of symposia held there since 1968. A surprisingly large number of industrial firms are now involved in the early exploratory work.

These symposia have already revealed that there is a great deal more to the subject of space industry than is immediately apparent. The results of the experimentation in *Skylab* confirmed this. The interest is so high at the moment that most industrial researchers do not feel that they can wait for *Spacelab* and the Space Shuttle in 1981; in the intervening years, space industrial experiments will be flown in a series of sounding rockets that will merely probe the edges of the space environment for a few minutes.

Most Americans think in terms of industrial products rather than industrial processes. Therefore, most of the revelations that have filled the press releases from these symposia, the *Skylab* experiments, etc. have concerned themselves with such mundane possibilities as true spherical ball bearings, high-purity crystals including super-sized diamonds, free casting of strange shapes, vaccine production, and formation of long high-strength filament materials for composite materials. But there is more to space industry than these early guesses at space products. To discover what these could be, we should have a very close look at each of the above characteristics of the space environment.

Mind you, what we will discuss here are just some of the characteristics that appear to be of industrial interest at this time.

And some of the potential processes in the space environment will use one or more of these characteristics combined. Together, they can produce some interesting synergistic relationships; that means that the combination turns out to be greater than the mere sum of the parts.

The specific characteristics of the space environment, insofar as they can be discussed as separate characteristics, are as follows:

1. Gravity-free environment:
This is perhaps the most unique characteristic of the space environment. It is also the most difficult one for us to understand or comprehend because we have grown up on the sur-

face of the Earth in a constant one-gravity field. This gives us a very distorted notion of physics and the way the rest of the universe operates. The discovery of total weightlessness is a totally new human experience.

We can come close to experiencing it in a tank of water if we ballast ourselves to achieve neutral bouyancy—adding just enough weight to ourselves so that we neither float nor sink. Although it is not a true weightless condition (our internal parts still "know" there is a gravity field), neutral bouyancy simulations of weightlessness have been extremely useful in teaching astronauts how to move and work in zero-g. (This is not an easy thing to do. Try removing or installing a drain or floodlight underwater in a swimming pool sometime without any help. You'll get an idea of the difficulties of learning how to work in weightlessness!)

It is also possible to achieve true weightlessness for periods of up to forty-five seconds by flying in a jet aircraft in a parabolic flight path. This is expensive, and therefore isn't something most of us have done. And the effect doesn't last very long.

Space is a totally strain-free, gravity-free environment, and it's there for as long as you wish. In fact, you must deliberately act to produce strain or acceleration.

Weightlessness is a natural condition of the universe. Being in a gravity field as we are is the really unusual condition.

Our lifelong experience tells us unconsciously that all liquids always need a container. Not so in weightlessness. Liquids become objects in their own right. They become things that can be moved and processed all by themselves in zero-g, perhaps not as solids, but with totally different methods than can possibly be used here on Earth.

In weightlessness, the ability to handle and process liquids *and* solids apart from any container as they float without support permits a high degree of control over contamination . . . or purity . . . or doping, depending on how you personally look at the business of combining one material with another. A solid or a liquid in weightlessness can literally be handled, moved, and processed without being touched by *anything.* This will really give industrial engineers the ability to regulate impurities!

Here is another important aspect of weightlessness that is not immediately grasped by our one-g minds: In weightlessness, density is no longer the dominant characteristic of a material. Density is a measure of weight or mass per unit volume. In weightlessness, the zero weight of all materials means that density is no longer as important in an industrial process. Density becomes apparent again when the material is accelerated, however.

Because density differences between materials no longer matter in weightlessness, ordinary convection heating currents in liquids and gases simply do not exist in zero-g. This will introduce some new factors into the heating and cooling of materials. Some of these factors will be desirable, while others will give engineers headaches. For example, since there can be no heat transfer by convection in a material, all heat transfer must then take place by conduction or radiation.

A fractional-distillation tower in a petroleum refinery simply won't work the same way out in space!

It is not necessary to put up with weightlessness in space if you have a process that demands a gravity field or can get by with a very similar acceleration field. If you put a centrifuge in the space factory, and if you are willing to put up with or design to compensate for the problems of gyro moments and torquing of the complex caused by the centrifuge when it starts up, runs, and stops, you will be able to create a controlled acceleration environment. A centrifuge permits you to create an acceleration environment to your specifications to suit your industrial process. You can generate pseudo-gravity in a centrifuge ranging from near-zero-g up to the acceleration that causes the centrifuge to come apart, its structural limit.

However, a centrifuge does not create the same pseudo-gravity of acceleration equally throughout its structure or even through its load. The closer you get to the hub of the centrifuge, the less the acceleration. Even in a long-armed centrifuge a hundred feet long, there will be a difference in the acceleration on the part of the load that is 100 feet from the hub and another part of the load that is 99 feet from the hub. This is known as an acceleration gradient. It is equivalent to a gravity gradient, something we cannot obtain much of here on Earth.

The smaller the centrifuge, the steeper the acceleration gradient. The faster the centrifuge turns, the greater the acceleration gradient.

This ability to create and sustain gravity-acceleration gradients to order is another totally new aspect of space industry. The ability to have one acceleration-gravity value in one part of an industrial process and a different one in another part opens all sorts of new possibilities, many of which we haven't even started to consider yet because we are still mentally chained to our one-gravity low-gradient terrestrial environment.

In an orbit close to a larger celestial body such as the Earth, it is possible in an orbiting satellite to obtain a condition of weightlessness combined with a gravity gradient. This is best shown digrammatically in the illustration on page 81. Point A on an orbiting satellite is farther from the center of the Earth than Point B. It is therefore subjected to a few thousandths of a gravity less gravitational attraction than Point B. This could be altered or controlled by changing the distance between Points A and B. The fact that Point B is subjected to a greater gravitational attraction than Point A keeps Points A and B in alignment, pointed toward the center of the larger celestial body. In turn, this causes internal stresses in the satellite structure because, although the entire satellite is orbiting as a single body at the same orbital velocity, Point A wants to move at a lower orbital velocity than Point B. This creates a "twisting" force on the satellite, and this force could get to be quite significant in a very large orbiting body. This should not be looked upon as a disadvantage; space industrial engineers will find some way to turn it into an advantage for some yet-unknown industrial process!

There is yet another aspect of the weightlessness condition

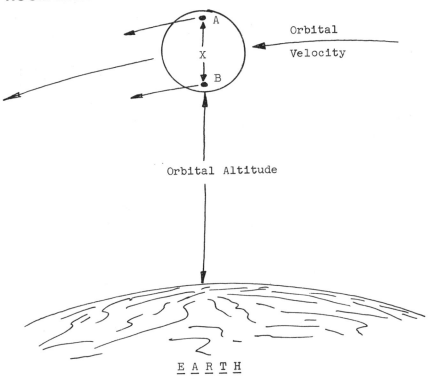

A gravity gradient exists even in a satellite. This produces a torque, or twist, between A and B. Eventually, A and B will line up pointing toward the center of the Earth below.

that we really don't have a gut feeling for as yet. In a strain-free, zero-gravity condition, the only forces acting on an object are its internal forces. Surface tension is an example of this in the case of liquids. Except the microscopic world of insects and large microorganisms, surface tension and other internal forces don't appear to be very strong. This is because on the Earth's surface they are masked or overshadowed by the very strong terrestrial gravity field. But when there is no strong external gravity field or other external force on an object, these internal forces suddenly become dominant.

In our one-gravity field, we find it very difficult to mentally separate weight from mass, and we usually totally ignore the aspect of mass called *inertia*. Unmasked by the absence of a gravity field, the inertia of a mass becomes one of its important and dominant properties.

And, finally, there is Coriolis force, named after the early nineteenth-century French engineer Gaspard Gustave de Coriolis. This is really not a true force, but an "effect" created in spinning centrifuges. It is something similar to the gravity gradient forces in a big satellite that were discussed above. It is best illustrated by the admonition "Don't shoot a game of pool in a rotating satellite!" There may be a pseudogravity gradient that keeps the cue ball on the surface of the table. But the table is turning with the rotating satellite. When you shoot the cue ball, it takes off in a straight line, but the table turns under it. It makes it appear as if the cue ball were traveling in a turning path. Satellite Fats is going to fleece a lot of gravel grippers from Earth before they catch on about Coriolis force. But it is an effect or pseudoforce that exists, and therefore something that can be used if needed.

All of these forces and effects will be useful in the normal weightless condition of space, even in rotating satellite-factories, because engineers will no longer be able to count on the gravity field to provide any motion. Any motion in the weightless condition must be provided by an applied outside force; see Sir Isaac Newton, whose Laws of Motion become realities. Many of us learned them in freshman physics classes, but we have yet to really live with them in their unvarnished states without a gravity field to distort them.

These strain-free conditions are already most exciting to researchers in space industry. They become even more so when combined, or "synergized," with the other factors of the space environment.

2. Matter-free environment:

A vacuum pump has been facetiously defined as a device that pumps nothing into a vacuum. And the only way to get a vacuum on the Earth's surface is to use this very expensive device, the vacuum pump.

The more vacuum you need, the more expensive the equipment and the greater the energy required to operate said equipment.

But at 500 kilometers orbital altitude (about 300 miles), there is a better natural vacuum all around than you could possibly get on Earth except with the finest and most expensive research-laboratory-type high-vacuum facilities.

The high vacuum of space is there for the taking.

As a matter of fact, there might even be some commercial value in space vacuum. It could be cheaply brought back to the ground in a space-going boxcar or a shuttle because it is coming down a gravity well. Eventually, it might be cheaper to bring it back from space than to make it on the ground; stranger things than this have happened in the history of technology and industry!

Even at that, the "hard vacuum" of outer space really isn't a very pure vacuum. We used to think that it was, before we actually measured it on the spot. In a near-Earth orbit, the vacuum of space still contains the tiniest microtraces of the Earth's atmosphere, and this contaminates the nothingness. In addition, the Earth orbits the Sun within the measurable solar atmosphere, a halo of hydrogen and other elements that stream out from the Sun. It is possible with modern vacuum-measuring instruments to actually put numbers on the near-Earth vacuum.

In freshman physics, vacuums used to be measured in centimeters of mercury, the height to which the somewhat negligible pressure of a vacuum would support a column of mercury. At Earth's normal sea-level atmospheric pressure, a mercury column is usually about 760 millimeters high, almost 30 inches. Today's laboratory vacuums are so good by comparison, however, that they are measured in "torr." The unit is named after Galileo's assistant, Evangelista Torricelli, who invented the barometer in 1643. One torr equals 1/760 of a "standard atmosphere," which is a column of mercury 760 millimeters high. Thus, a torr is also one millimeter (about 0.040 inch) of mercury column.

The vacuum inside your television picture tube is not very

good by today's high-vacuum standards. It is about a millionth of a torr, written 0.000001 torr or 10^{-6} torr.'

In near-Earth orbit at 500 kilometer altitude, the pressure or vacuum (depending on how you want to look at it) is about 8.22×10^{-9} torr (or 0.00000000822 torr).

Although this is a pretty good vacuum by earthbound standards, it is possible to get an even better vacuum very easily.

A satellite moving in near-Earth orbit creates a wake behind it in the ultrathin atmosphere, just as a boat does when moving through the water. The satellite velocity is so high (measured in miles per second) and the vacuum of space so thin that this satellite wake has a very low pressure indeed. For a satellite 10 meters (about 30 feet) in diameter, the inner wake region can have a vacuum as high as 3.69×10^{-15} torr (0.00000000000000369 torr). And it is possible to get this for nothing. It's there in the satellite wake.

Starting from the next-to-nothingness of this very good vacuum, it is possible to create various atmospheres to your exact specifications. Name the pressure and composition, and it is possible to get it readily by introducing whatever you

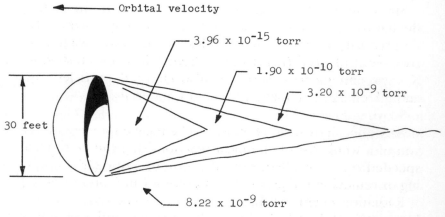

Orbital velocity

3.96 x 10^{-15} torr

1.90 x 10^{-10} torr

3.20 x 10^{-9} torr

30 feet

8.22 x 10^{-9} torr

Orbital altitude = 300 miles

Even though a very good vacuum exists at orbital altitudes, even higher vacuums exist in the wake of a satellite.

want in the quantities you want into this hard vacuum. You are limited only by the structural strength of your container; it will hold as much pressure as you design and build it for.

The matter-free environment can be very highly controlled in this manner since there is so little there to start with. Thus, an industrial engineer can introduce whatever he wants into a container of this hard vacuum to suit his industrial process.

Pressure, vacuum, and high-purity atmospheres are already widely used in many earthbound industrial processes today. They are part of our industrial world. Although industrial engineers of the future won't have to learn a great deal about this, they will have available some capability that they did not have before—and economically, too.

3. Controlled radiation environment:

The Earth's atmosphere is an excellent filter. It strains out certain parts of the electromagnetic energy that streams down on Earth from the Sun and from interplanetary space. There are only certain "windows" in this atmosphere. The natural radiation that finally impinges upon the Earth's surface is only a feeble remnant of what exists above 200 miles altitude.

Space industries are going to have available to them all of the natural radiation emanating from our neighborhood fusion reactor, the Sun. This includes not only radio waves, microwaves, infrared (heat) waves, light, ultraviolet radiation, X rays, and gamma rays, but also charged particles with enough energy to ionize a number of materials.

So what good is this?

Any good industrial chemist or chemical engineer will tell you of a whole list of chemical reactions that can be started, speeded up, controlled, slowed down, or stopped by applying or removing radiation of various types and intensities.

Radiation chemistry has always been somewhat retarded here on Earth by a lack of suitable, effective, and economical radiation sources in certain parts of the electromagnetic spectrum. There are also definite limitations on the amount of certain types of radiation that can be obtained from earth-bound sources.

Not so in space. There's old Sol, radiating away energetically in a wide swath across the electromagnetic spectrum.

To get more of what the Sun produces, an engineer can concentrate or focus certain parts of the electromagnetic spectrum such as light. Or he can locate his industrial operation closer to the Sun.

There is radiation aplenty available out there. It is a form of energy that will be useful in industrial processes once space engineers learn how to handle that plentiful supply.

4. Wide-range temperature environment:

There has been much written about the cold black depths of space or the searing heat from the Sun on a spacecraft. Space itself, being a vacuum, has no real temperature as we have grown up to understand it. We associate temperature with heat energy capacity. But, in space, we must mentally separate temperature—which is a measure of the vibrational speed of molecules and atoms—and heat, which is the quantity of energy available from the vibrating and speeding molecules and atoms.

Temperature and heat are two common factors in many earthbound terrestrial industries. In some cases, we have run up against the problem of heat pollution, of inadequate heat sinks into which to dispose of waste energy.

The space environment offers industry a wide variety of temperatures.

Between the temperatures of $-40°$ C $+ 3000°$ C, it is still reasonably economical and easy to conduct industrial operations on Earth's surface. But if low temperatures are required, the mechanisms for getting them become complex and expensive. In fact, it gets downright impossible when temperatures near absolute zero are required.

High temperature regimes above $-3000°$ C are very difficult and expensive to obtain and maintain. Furthermore, they will become more and more expensive, more difficult, and more antisocial as Earth's supply of fossil fuels runs out, as our heat sinks become even more inadequate, and as our antipollution laws subsequently become stiffer.

We will probably be able to continue with some of the high-temperature industrial operations and processes here on Earth for a time if we do not overload the ability of the ecology to handle the waste heat beyond the point where it can recover—and provided there is a source of heat energy available.

On the other hand, space industries will have available to them a range of temperatures running all the way from nearly absolute zero ($-273°$ C, give or take a fraction of a degree), up to temperatures of a million degrees or more (C or F—it doesn't make any difference at that level!) in the solar corona or in a solar energy collector. With a nearby star radiating 80,000 horsepower per square yard of its surface continuously, there is not only plenty of high temperature available, but also lots of heat energy as well.

Ultra-low temperatures can be obtained by shading an object from the Sun. Its temperature will soon drop to about $-269°$ C or thereabouts, which is the normal temperature of interstellar space. And there are techniques that can be used to get nearly to absolute zero. And they aren't laboratory-level technology, not in space.

In space, it is also possible to sustain a given temperature for a very long period of time with little energy input. Space is literally a great big vacuum bottle.

Very low temperatures are fascinating to consider for industrial processes. They make possible a whole new realm of processes and techniques—such as common use of superconductors.

5. *Wide-range energy density environment:*
Many of today's earthbound industrial processes require high energy densities—i.e., a very high concentration of energy. Melting, smelting, vaporizing, solidifying, freezing, subliming, fractionating, and a host of other processes require that energy be put into or dumped out of the process at various rates.

In the space environment, it is not immediately obvious that comparable levels of energy density, both high and low, can exist. In fact, they exceed anything we can achieve here on Earth. Space is usually considered as a place with very low energy density. True, most of it is this way, especially out between the stars. But there are nodes of extremely high energy density, such as a star.

This provides us with a variety of energy levels to work with and a wide variety of possible energy density gradients, much better than anything on the Earth's surface.

It's all a matter of using what's out there—and there is a lot of it—in a way that accomplishes what you want.

As a matter of fact, this can be said for all of the space characteristics we have discussed above.

We have been guilty of thinking of space as a place with a view. We can now start thinking of it as a place we can work in, too.

It's a rather nifty environment for a lot of industrial processes we now have and a lot we will develop once we become more familiar with space.

7

Made in the Solar System—I

ABOUT THE ONLY people who are deeply interested in industrial processes are industrial engineers. Marketing men try to figure out what can be sold and how much of it. They tell managers and executives, who then talk with the financial people to find out if they can get enough money to do it. They finally give the engineers some money and tell them, "Let's have the product, and don't bother me too much with the details."

Granted that this is perhaps an overgeneralized statement, nevertheless it is true in more cases than not.

But the whole ball game in space industry is processes.

Because with the processes we may be able to use in space, we will probably be able to make nearly everything we can make here on Earth.

Furthermore, we may be able to deliver the goods at or near the same price (at a lower price, we hope).

And without tapping terrestrial energy or material sources. Or polluting the environment.

But, most important, we may be able to develop some new and better processes that will make existing products cheaper or better. And we will most certainly use the space environment to develop totally new processes that can be used *only* in space for making totally new products that don't even exist today.

This should take no one by surprise. More than 50 percent

of the products that are available today did not exist twenty
years ago. Right here on Earth, too.

Because of this, we *must* discuss processes. I have some
ideas on how these processes might be used to produce some
existing or new products. But you will have a different perspective than I have, and it is a good bet that you will put
two and two together a little bit differently, based on your
own experience, to think up some products on your own.

And high time, too, I might add. There is considerable
consternation among industrial research and development
people today over the fact that the number of patents being
issued and the rate at which patent applications are coming
in have decreased sharply in the last ten years. Whether this
is due to the patent law situation, to the advisability of keeping trade secrets in lieu of patent rights, or a general decrease in the number of new products being developed is
pretty much beside the point. And beyond the scope of discussion here.

If we have some way to get into and around space, if there
are raw materials there capable of being used, if there is energy enough there, and if the space environment can be
viewed as useful, we now must consider what we can do with
all of these things.

First of all, how are we going to use the energy that is available out there?

We can use it directly as heat, of course, as was pointed out
previously. But this is rather limited in its application. At the
moment, electricity appears to be the most efficient means
for utilizing the energy sources of space—solar, nuclear, and
chemical.

Electricity is easily generated from all three sources using
known techniques. Electricity is easy to handle, meaning that
we know how to generate, control, transmit, measure, switch,
and use it. We can put it into a "bucket" called a battery or
the hydrogen-oxygen dissociation of water. We can then
store it and call it forth like some genie of old to do our bid-
ding when we want it.

NASA has studied several promising electrical heat

sources for possible use in space industrial processes. These are:

1. Induction heating
2. Electron beam gun heating
3. Electron beam plasma gun heating
4. Laser
5. Electric arc
6. Electric resistance heating
7. Ultrasonics
8. Microwave heating

Some of these methods can easily be used in a vacuum. In fact, some of them require a vacuum in order to work at all. Others require an atmosphere, often a very specialized atmosphere. But, as we have previously discussed, hard vacuum as well as controlled atmospheres are part of the general space environment or industrially possible in space.

Electric arc heating is one of those methods that will not work in a vacuum. A spark or arc will not form or sustain itself in a vacuum. It requires an atmosphere that can be ionized by an intense electric field before the arc is struck and that can remain ionized by the arc's energy itself afterward.

You might think that a laser would be an excellent industrial heat source for use in space. The laser is a very new and therefore a very romantic technical device, but it has its limitations. According to unclassified sources of information, the best efficiency thus far attained by high-powered carbon-dioxide lasers is a roaring 14 percent, which means that 86 percent of the energy originally put into the laser is waste energy. And it must be gotten rid of or handled somehow. As a space industrial heating source, the laser may be useful for specialized applications, but not for general heating applications. This statement is subject to modification or downright retraction, however, because, as stated above, the information is based on *unclassified* data. The most powerful and efficient lasers at the time of this writing are those being investigated for weapons applications. In other words, ray guns. It is known that the Department of Defense has developed some rather powerful lasers. One of these is reported to have punched a hole through one inch of armor plate at a distance of one mile; the concentrated heat energy burned its way through the armor steel in a fraction of a second. It is also reported that unmanned drone aircraft have been brought down with laser weapons. Someone therefore has worked out some of the efficiency problems. So please do not write off the laser—yet. The defense-classified laser data may one day be rediscovered independently in some industrial research lab whose security is usually tighter than that surrounding a weapons research lab! And the laser may show up on the industrial scene in space in a big way.

The category of "low efficiency" also applies to ultrasonic heating sources. Efficiency in energy conversion and utilization is something that is going to be important in space industry for a different reason than it is on the ground. Because of the ready availability of energy in space, the problem becomes one of being able to get rid of the waste energy before it melts the space factory rather than seeking efficiency for economic factors. Ultrasonic heating utilizes the high-frequency vibration of a material to produce frictional heating; by its very nature, it is not efficient as a process. It may be even less efficient in zero-g.

Microwave radiation heating suffers from poor heat efficiency and poor weight efficiency at this time. Essentially, this is the heating method used by domestic microwave ovens. Microwave electromagnetic radiation generated by a special electronic device excites the water molecule in foods, for example. Thus, it is molecularly sensitive at relatively low beam powers. Yes, if you have enough beam power—megawatts concentrated in a beam with high energy density—such as in a high-power missile tracking radar, you can heat a lot of things. A ball of steel wool tossed into the beam of such a missile tracking radar will be vaporized. But it takes a whopping lot of equipment to do the trick—at this point in time, anyway.

The electron beam gun may turn out to be a very good industrial heating source for space industry. The picture tube of a TV set contains an electron beam gun; it is small and low-powered; the beam of electrons it puts out does not appreciably heat the face of the TV picture tube. But electron beam guns can be made bigger. They require a good vacuum for their operation (and there is a lot of that in space). And the electron beam thus produced can be directed and focused on a target with well-known techniques. As a local heater, it works very well already in industry.

The electron beam plasma gun is a more complex version of this. The electron beam is set up in a plasma, an envelope of heated, electrically charged gas. As a localized heater, it is also very good and currently used in earthbound industry.

Induction heating comes in with a score of 70 percent efficiency. However, it requires that the material to be heated be electrically conductive because it uses a powerful electromagnetic field to set up eddy currents in the heated material. It is possible to get very high temperature gradients with induction heating. In addition, if the heated material is magnetic in nature, induction heating techniques can also be used to suspend or levitate the material without supports.

The very best heat source for space industrial purposes in terms of thermal efficiency, weight efficiency, cleanliness, and adaptability to the space environment is electric resistance heating. The good old home electric stove uses this form of heating. Electric energy is directly converted into heat energy by passing it through an electrically resistive material—a Calrod unit, if you will. This produces heat directly from electric energy. It can be applied locally or generally, depending upon the shape and size of the heating unit. It will probably turn out to be the most widely used heating method in space industry.

Please note that all the heating methods briefly discussed above are characterized by the absence of a flame created by the combustion of a fuel and an oxidizer.

This is not because a flame heating method requires both a fuel and an oxidizer.

It is because the weightless environment of space does not permit the existence of a flame as we know it.

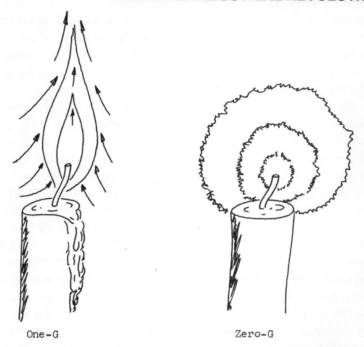

One-G   Zero-G

The behavior of a simple flame is greatly different in zero-g from that which we are used to in our one-g earthly environment. On Earth, less-dense gases rise and pull in fresh oxygen for the flame. In zero-g, there are no convection air currents formed, and the flame surrounds itself with combustion products.

A flame in zero-g is not a heating method; it is quite different and is a chemical processing device instead.

The familiar candle flame in our familiar one-g field is shown diagrammatically above. It is actually a very complex, high-speed chemical process. The flame results from burning the paraffin hydrocarbon material of the candle in the oxygen of the surrounding air. Once initiated, a candle flame is a steady-state process, a chemical chain reaction that provides enough energy to keep itself going so long as there is fuel and oxidizer available to it.

Several chemical events occur very rapidly in this flame— or in any other flame, for that matter; we are considering a candle flame here because of simplicity of illustration—and combustion experts are still not exactly certain of what happens where and when. The simple flame turns out to be another common and apparently simple phenomenon like the electrical transformer that turns out to be very complex and very difficult to understand.

Briefly, here's what the experts think really happens in the common, household variety of candle flame:

The solid paraffin hydrocarbon material of the candle body is melted into the liquid state by radiant heat from the flame itself. This liquid is then transported to the flame by the capillary action of the candle wick. Once delivered to the flame site in liquid form, the paraffin is then vaporized by heat radiated from the flame. These paraffin gases then react with the surrounding atmospheric oxygen gas. The resulting oxidation reaction is exothermic; it releases heat. This heat continues to drive the reaction. Various gases are formed by the combustion process; these include carbon dioxide, carbon monoxide, water vapor, and numerous minor combustion products ranging in chemical complexity from simple diatomic hydrogen to large organic molecules.

Specialists in flame structure and combustion processes believe that many of these combustion product compounds are formed through various stages of reaction that involve production of free atoms and free radicals. In other words, momentarily in various regions of the flame, atoms exist all by themselves or exist in combination with others in a radical or chemically ionized state.

The resultant low-density hot combustion end-product gases rise up from the flame zone because they are less dense than the surrounding atmosphere. This permits more atmospheric oxygen to enter the flame zone and take part in the continuing chemical process.

Although flame combustion is therefore a steady-state process when viewed grossly, it is also a very complex, step-by-step chemical process with many intermediate steps. Fuel flow and atmospheric oxygen flow into the system can be equated with gas flow out plus heat energy liberated.

If the flame process occurs in a gravity field!

In weightlessness, something else happens.

If the candle and flame are placed in a zero-g environment, the lower-density combustion gases do not rise away from the flame zone. They can't. In the absence of a gravity field, less-dense gases cannot rise. Therefore, fresh atmospheric oxygen cannot move into the flame process zone to perpetuate the combustion process.

A flame in zero-g is like no flame you could ever see on Earth.

A simplified zero-g flame shortly after ignition is shown on page 94.

We know what a zero-g flame looks like and generally how it behaves because of a number of studies made in jet aircraft flying in what are known as parabolic trajectories. It is possible to attain 20 to 30 seconds of weightlessness in a jet airplane flying on a parabolic trajectory. The plane dives to build up speed, then pulls up to fly in a great, long arc through the sky—up to the top of the arc and then down again, all under careful control so that the plane falls up and down again in a state of weightlessness—which is also called "free fall." Everything in the plane falls up and down with the plane and is therefore in weightlessness. The technique was widely used to train astronauts in zero-g maneuvers.

The studies made on flames in weightlessness in jet aircraft have shown some interesting and, at first, confusing results. Much of the information came from high-speed motion pictures taken of the flame in zero-g.

Initially, the zero-g flame builds up to maximum size and brilliance very quickly.

Then, just as quickly, it recedes and darkens.

In actuality, none of the zero-g flames studied thus far have achieved the full development shown in the idealized sketch because the ignition method localized the burning to a few spots. In addition, the combustion products from the flame tended to subdue the flame before the fuel-wick system could be totally enveloped by the cloud of gases from the combustion.

But, although the flame itself appeared to go out, *the zero-g flame process did not stop!*

When the acceleration of gravity returned at the end of the parabolic flight trajectory, the flame reappeared!

This was totally unsuspected. One would tend to think that the zero-g flame would go out because the hot gases could not rise away from the flame zone and allow fresh atmospheric oxygen to come in and continue the combustion process. The zero-g flame is the sort of totally unsuspected and highly serendipitous sort of behavior that we should expect. After all, we are children of a gravity field, which distorts the real universe. And it is probably the first of the new space-based industrial processes.

A simple flame in zero-g is not a heating source; this will have to be carried out with flameless electricity and solar energy. A flame becomes a chemical processing system, to wit:

Naturally, the strange behavior of the flame in zero-g immediately invokes scientists to form hypotheses in an attempt to explain or account for what happened.

One of these hypotheses goes as follows:

The gas that is formed by the fuel by heat energy reacts chemically with the atmospheric oxygen in the classic combustion reaction process. In a gravity field, this progresses in a steady-state fashion with a constant ratio of fuel to air, a steady mixture ratio. But, in the weightless condition, the zero-g flame experiences a constantly changing mixture ratio. It goes from very lean (lots of oxygen and little fuel) to very rich (little oxygen and lots of fuel) in a very short period of time as the surrounding oxygen is used up in the process. This process uses up the oxygen much faster than ordinary diffusion processes can replace the oxygen. Oxygen starvation of the flame therefore takes place. The flame immediately begins to cool as the mixture gets richer.

The result is a blanket of fuel-rich flammable gas surrounding a reservoir of molten fuel that is in turn covered by layers of both solid and gaseous combustion products. These combustion products are the ones normally present for a split second in an earthbound flame; but, in the zero-g flame, they are "frozen" in their intermediate condition by the rapid cooling of the flame region. This blanket of gas and combustion products probably contains both free atoms and free radicals, plus a lot of very fuel-rich components. Plus the heat energy that has not been able to leave the region by convection currents. In other words, the heat energy stays in the system, leaving only by means of conduction through the blanket of gases, etc. Or it may leave by direct radiation from the low-temperature corona that surrounds the system.

At this point in time the zero-g flame can best be described as *dormant.*

When convection is renewed either by acceleration of the system or by fan-induced air currents, the gaseous corona is cleared away; oxygen is again provided to the system; the paralyzed combustion products begin to react again because of the latent heat that remains, and the flame is renewed and resumes.

How long will a zero-g flame remain dormant?

The aircraft tests provided only enough weightlessness for a maximum of 12 seconds "float" for the zero-g flame, what with the time required to get it started, etc. This is plenty of time to study something, however. And to sample the gases and compounds within the dormant flame. And to measure the hell out of it.

The flames remained dormant during 12-second zero-g floats.

We don't know how much longer they will remain dormant.

But it doesn't make any difference.

Ten to twelve seconds is not only plenty of time to study a zero-g flame with modern, high-speed instrumentation, it is also plenty of time for a hypothetical, yet-to-be-devised industrial chemical process to get all sorts of interesting products from a zero-g flame.

By tapping a zero-g flame, it may be possible to obtain some chemicals that we cannot obtain in any other way.

We can't tap them on Earth in a one-g field because they don't hang around that long.

But in zero-g, where they can be frozen in chemical paralysis for seconds at a time in various regions of the zero-g flame zone, there is plenty of time to get in there, tap them, drain them off, combine them, and come out with some wild radicals and other chemical curiosities.

We spoke of a candle flame only in our discussion. But the zero-g flame system can be expanded to include flames of any type, fueled by any of thousands upon thousands of fuel-like chemical compounds, and reacted with thousands upon thousands of different types of oxidizing compounds. We don't yet know what we can make with the zero-g flame system alone; it is still in its barest infancy as an industrial process. And it will require the weightlessness of space for any real industrial research to begin.

We rather sneaked in the back door of space processes here, starting with heating methods and progressing into consideration of a terrestrial heating method that becomes a chemical processing system in space. We are going to run into other amazing and serendipitous discoveries in the space environment. Obviously, because of the one-g mental orientation we've all grown up with, we couldn't possibly suspect all the zero-g possibilities of things that are rather commonplace phenomena here. The reason is simple: We have a distorted notion of the way the universe works because we grew up on a planetary surface.

Most of the new space industry discoveries are yet to be made, but most of them will probably be very elegant in their simplicity and will cause us to exclaim, "Why, of course! Why didn't we think of it before?" The simple flame in weightlessness and the complex chemicals that are available from it is an example of this. It is also an example of the sort of things to come from space industry that will bear the label MADE IN THE SOLAR SYSTEM—because that is the only place where they can be made.

CHAPTER

8

Made in the Solar System–II

WHEN WE THINK about things that can be made in space, we do not necessarily have to base our speculations on what we might or might not be able to discover to be possible once we get out there to find out. We don't have to withhold early industrial research until we get the *Spacelab* and the Space Shuttle to try them out *in situ.*

The people who are already interested in space industry are doing their homework in advance. They have already identified a number of industrial processes that appear to be quite viable, unique, and economical in the space environment.

As you might expect, NASA is interested in space industry—"space processing," in their lexicon. But the list of corporations that have taken part in space industry symposia, sponsored or conducted research in the area of space processing, and already committed to keeping up with the field (probably as a hedge against the time when Space Shuttle and its progeny makes it more economical) is long and impressive:

American Optical Company, General Electric, Grumman, Tyco Labs, Rockwell International, Revere Copper & Brass, Martin Marietta, Westinghouse Electric, General Dynamics, Lockheed, Bendix Corporation, Boeing, Chrysler, E. I. du Pont de Nemours, General Motors, IBM, National Lead Company, Owens-Illinois, Reynolds Metals Co., Union Carbide Corporation, and Western Electric are only a part of the list that is growing every day. True, most of them are very large corporations that can afford to set aside a little bit of funding to hedge their bets. But you can also be certain of the fact that they wouldn't support this diversion of the stockholders' money unless there was very good reason to believe that there would be a payoff someday.

Since 1968 NASA and some of the firms listed above have hammered out rationales, justifications, and advanced development targets for space industry.

They have already identified a number of industrial processes that appear to be quite workable, unique, and economical in the space environment.

The list on pages 114–16 shows in outline form 14 different generalized industrial processing areas amenable to the space environment. These include 43 different subheads that are distinct processes themselves. Each of them can be combined with others on the list to create more complex processes. These combinations may be synergistic, i.e., they may produce products that could not be produced by either of the two or more combined processes acting alone.

The total number of combined industrial processes that could be obtained by combining these 43 items in all different combinations is staggering. The number is truly astronomical: 1.642×10^{53}.

That's 1642 followed by 50 zeros!

We don't even have a word for a number that big—it's about half a googol, for those who are math sharks, or a number slightly bigger than all the stars in the entire universe. Obviously, there are going to be a lot of industrial processes that we can try out in space. Some of these combinations will turn out to be impossible, silly, and ridiculous. Others won't work at all, as we will probably discover after trying them. Some of them won't produce anything. But even if only one percent of them work and are useful, it gives us 1642 followed by 48 zeros of industrial processes.

In other words, we'll never have the chance to try them out. There aren't that many people, and there won't be that many people born before the universe comes to an end 50 billion years or so in the future. So even if only one out of a trillion combinations ends up being possible, praçtical, and economical, there is still more to do than time to do it—and so there are many different approaches to space processing.

But a lot has been accomplished thus far to check out the most promising near-term space processing possibilities. We can already begin to look toward the profitable operation of some processes in the near future—within the next decade.

Although the first technical symposium on space processing was held in Huntsville, Alabama, at the NASA George C. Marshall Space Flight Center in November, 1968, and the first in-space processing experiment took place aboard *Soyuz 6* when Valery Kubasov performed experiments in welding with the "Vulkan" apparatus in October, 1969—there has been the sort of quiet progress in the "back rooms" of industry that characterizes most early development work. The *Skylab* flights provided the first long-duration, planned opportunity to test some space processing theories. In the interim between *Skylab* and Space Shuttle, NASA will be flying dozens of small, inexpensive-sounding rockets to provide several minutes of zero-g for experimenters who just can't wait for the Shuttle. Space industrial conditions and environments already have some engineers wildly excited over what can be done.

For example, separation and purification of materials are techniques that are widely used in earth-based industry. Nearly all smelting and refining operations include these elements. In the space environment, they take on wholly new aspects. On Earth, electrophoresis is a separation technique of limited usefulness; in space, it becomes a major industrial operation.

Electrophoresis uses electric field gradients to separate molecules of different types. Electric fields cause molecules and particles to move and separate. As the various components are separated, they are removed from the solutions. Gamma globulin, the blood component used as a specific treatment for several diseases, was first identified in this manner.

But a gravitational field causes settling of the solutions and sets up convection currents that make warmer components of the solution rise. This really fouls up the electrophoresis technique. On Earth, liquids undergoing separation by electrophoresis must be confined in thin films to prevent convection currents. This naturally limits the use of the electrophoresis technique to small batches and laboratory testing.

Convection currents are caused, of course, by less dense materials rising in a solution. In zero-g, there can be no convection currents.

Therefore, the electrophoresis technique is not necessarily limited in size and output quantity in zero-g.

Two experimental demonstrations of space electrophoresis technique were made aboard *Apollo* and *Skylab* flights. They were highly successful. They involved the highly efficient, straightforward, and simple fluid electrophoresis technique; on Earth, we have been confined to using the smaller and more complex techniques of paper electrophoresis, column electrophoresis, and electrochromatography. The pharmaceutical industry has spent millions and millions of dollars over the years in a constant program of research to develop electrophoresis and other techniques to separate, refine, and purify vaccines, serums, blood fractions, enzymes, and other biological products. In some of these, the slightest trace of impurity can cause dangerous side effects or drastically reduce the effectiveness of the biological preparation. Some highly effective experimental vaccines cannot be made in sufficient quantity because of lack of high-quality refining processes.

But in the zero-g environment of near-Earth orbit—right over our heads—we could set up a space pharmaceutical factory using Space Shuttle that would permit the volume manufacture of these high-quality vaccines.

Right now, with what we know as this is being written, we could process the ten most commonly used vaccines in space in a quantity of about one ton per year—the usage level if all other nations in the world were that of the United States—with an estimated value of over $1 billion per year. We can do this with the currently planned Space Shuttle alone.

And in the pharmaceutical research laboratory adjoining

this zero-g production facility, researchers are already looking forward to using the improved techniques to help find the cure for the common cold, influenza, and a host of other diseases and respiratory ailments.

And electrophoresis is just *one* industrial process that is possible in the space environment.

Process produces products, and vaccines are only one of the potential space products that will change our lives as drastically as the Hall Process for refining aluminum has changed it.

The lack of convection currents caused by density differences in materials is just one aspect of the weightless condition of space that is just beginning to draw attention to itself. Convection is such a strong factor here on Earth that it masks and hides other forces that are weaker in comparison. Such forces include surface tension. In the space environment, the absence of gravity and convection currents makes these formerly "weak" forces very prominent indeed. They become, by themselves, forces that can be used in industrial processing.

There are two prime examples of surface tension phenomena that are currently of interest to space processing investigators.

Very few people other than physicists have heard of Marangoni Flow, but many people have witnessed earthbound examples of it.

Marangoni Flow is fluid flow caused by a surface tension gradient.

Surface tension may best be simply described as a gross molecular adhesion force within a fluid at the interface between the fluid and another fluid. The strength of the surface tension force depends upon both the temperature of the fluid and its chemical composition.

If a free liquid surface has a surface tension gradient because of a difference of temperature or chemical concentration between different points, the fluid will flow on the surface from an area of low surface tension force to an area of higher surface tension force.

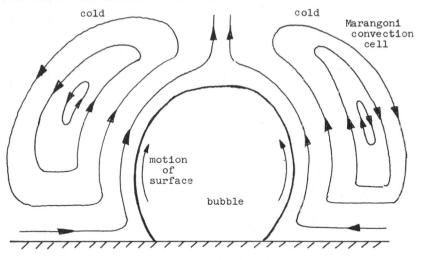

Marangoni Flow is a surface-tension phenomenon caused by the movement of a fluid surface from an area of high surface tension to one of low surface tension. Surface-tension gradients can be caused by a difference in temperature of chemical concentration.

This sounds complicated and difficult to understand, but it is so only because of our "one-g" orientation caused by growing up in the Earth's gravity field. Marangoni Flow will be a commonplace thing to future space dwellers who grow up in weightlessness.

Actually, many of us have seen Marangoni Flow here on Earth. Sometimes when conditions are just right, you can see "tear drops" forming inside of a wineglass that is partly filled with chilled wine. Swirl the wine until the inside of the glass above the wine level is wet by the wine. The evaporation of the alcohol in the wine chills this layer of wine that is on the inside of the glass. The difference in temperature causes this wine layer to have a higher surface tension force than the warmer wine in the lower part of the glass. So the surface tension difference draws liquid from the bulk lower portion until a "tear" is formed. When the tear becomes large enough for gravity forces to prevail over the surface tension forces in the immediate vicinity of the drop, the tear runs back down the inside of the glass and into the bulk of the wine supply again.

What can we possibly do in space with something as apparently academic as Marangoni Flow, today's wine-tasting table trick and strictly a laboratory curiosity?

Anything that can cause material to move is an important factor in an industrial process. Anything that can separate because of a difference in materials or physical conditions is fair game for industrial engineers. In the absence of gravity in space, Marangoni Flow can be used to separate fluids of differing surface tensions; these differences in surface tensions can be caused by differences in temperature brought about by different specific heats or by simple differences in chemical composition. Marangoni Flow can be used to pump fluids from one place to another by the application of heat to one part of the process, thereby creating a heat-driven pump with no moving parts that will work in zero-g. Marangoni Flow can be used to cause fluids to move under low pressure conditions. Do not fall prey to thinking that Marangoni Flow is applicable only to room temperature fluids such as the wine of our example; Marangoni Flow will also work with molten steel, which is a fluid, or with liquid hydrogen, which is also a fluid. It is unlikely to work with two gases because of rapid diffusion that takes place at the interface between two gases, but there might be future space industrial applications in which a great deal of gas is used and pumped by Marangoni Flow.

Other applications of Marangoni Flow will be developed as space industrial engineers—some of whom are already born—become acclimated to the space environment and start to shed their one-g orientation. It's a process waiting for an application. Fractional distillation was once just a process, a laboratory stunt for separating liquids/gases; today it's the basic process of hundreds of petroleum refineries processing millions of gallons of petrochemicals every day.

Another surface tension gradient phenomenon is a variation of Marangoni Flow called Bénard Flow after the scientist who first observed it in 1900.

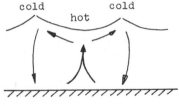

BENARD FLOW

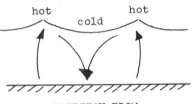

RAYLEIGH FLOW

Rayleigh Flow is the well-known earthbound formation of convection cells by density gradients where the rising of low-density fluid lifts the liquid surface as at right. But in zero-g, the only convective heating cells that will form are the result of Bénard Flow, a surface-tension phenomenon where liquid flows from areas of low surface tension to areas of high surface tension and flows faster on the surface than within a fluid. Thus, Bénard Flow lifts the fluid surface over the cold portions of the convection cell because surface tension increases with a decrease in temperature. It doesn't look right, but it worked in the *Apollo* and *Skylab* experiments.

Bénard Flow will produce a semiconvection cell in a film of liquid in the absence of a gravity field or even against a gravity field.

Take a thin layer of fluid on a metal plate. In zero-g, the fluid will adhere to the plate by surface tension forces and will, in fact, try to cover all the surfaces of the plate with a film of reasonably uniform thickness. Stick the film-covered plate in a cool atmosphere and heat the plate. In effect, this heats the film on one side and cools it on the other.

Normally, you would expect to get ordinary, commonplace, run-of-the-mill convection cells called Rayleigh convection cells. This is what we experience in a coffee pot in one-g on Planet Earth. Heated areas form in the liquid and, being warmer and less dense, therefore rise in the fluid toward the cool surface. At the cool surface, they lose their heat and become more dense, creating descending cells in the fluid.

Not so with Bénard Flow, which is a surface tension gradient phenomenon and, in a one-g field, is overpowered by the normal density-caused Rayleigh convection cells. The fluid within the thin film flows from the cooler region of low surface tension to the warmer region of higher surface tension. It works backward from Rayleigh convection cells. And it works only in a thin film of fluid, not in a volume of fluid.

Again, Bénard Flow is a process phenomenon. Motion is produced by an energy gradient. Therefore, it can be controlled and put to work separating and transporting fluid materials.

The initial space experiments with Bénard Flow are a matter of history. The first experiment took place on *Apollo 14;* accelerations of about 0.0001 g due to spacecraft motion produced some questionable results, although both steady-state and oscillating convection cells owing to Bénard Flow were observed. On *Apollo 17,* spacecraft motions were held to accelerations of less than 0.000001 g, and better data was obtained. These preliminary tests in *Apollo* flights were followed by more complex experiments in diffusion and crystal growth conducted aboard *Skylab.*

In order for Bénard Flow to take place, however, the fluid must "wet" the warm surface. If it does, heat can be extracted from the surface more readily. If it doesn't, the non-wetting film doesn't get as hot because heat will not bridge the gap between plate and non-wetting fluid as easily.

We see this sort or behavior here on Earth every day. Water wets most all materials, and we therefore see a meniscus, as shown on page 109. When a material such as mercury does not wet a glass container, another sort of meniscus is formed, as shown in the same illustration.

Because some materials will wet a surface and others will not, the phenomenon can be used in space for separation of materials.

Because some materials will wet a surface and others will not, causing a difference in heat transfer rate, we can use this phenomenon in space for differential heating, sort of a zero-g type of fractional distillation, if you will.

We can also use the wetting and non-wetting phenomenon in space as a way to (1) distribute the fluid uniformly over a surface if wetting takes place or (2) keep the fluid away from the surface if non-wetting takes place.

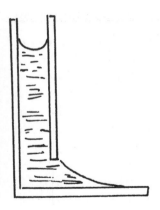

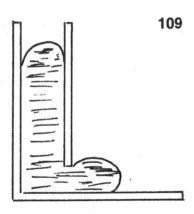

water   mercury

(wetting)   (non-wetting)

Examples of the shape of a meniscus for liquids that exhibit both
wetting and non-wetting characteristics.

Thus, we can pump one fluid without moving another
one.

These surface tension effects are fascinating. They are as
important to us in space as they are on Earth to insects walking on the surface of water—or to dirt particles being washed
out of fabrics by soaps and detergents, which operate on surface tension effects, too. Certainly one cannot say that the
surface tension effects of soaps and detergents are not important industrial tools here on Earth, to say nothing of valuable household tools as well. In space, where the lack of a
gravity field allows surface tension effects to come to the fore,
they will be more important than ever in making space products.

Separation techniques are not confined to those utilizing
surface tension effects. In the absence of gravity where these
surface tension effects become prominent because of the lack
of convection due to density differences, it is also possible to
create density differences made to order for separation.

On Earth, any medical laboratory these days has a little
centrifuge for separating the various components of blood.
A centrifuge creates a very high acceleration that separates

materials of different densities. The materials with higher densities collect on the outer end of the diameter of the centrifuge while those of lower density collect toward the center of rotation.

Although we have weightlessness as a matter of course in space, we can also create accelerations and pseudogravity to order by building and operating a centrifuge. Some of these space processing centrifuges will be small. Others will be hundreds of feet in diameter. We will use them in the same manner as we currently use small centrifuges in medical labs: for separation of materials of differing densities. But we'll use them in space to create the density-separation process that normally exists on Earth and is totally absent in space.

And we will use centrifuges for another purpose, too: to create acceleration gradients.

This is because the acceleration caused by the rotation of a centrifuge depends on how far you are from the center of rotation. On the end of the centrifuge arm or wheel, the acceleration caused by centrifugal force is greatest. As you move in toward the center of rotation, the acceleration becomes less and less. In free-fall, you would be in a weightless condition at the center of rotation. Thus, there is a difference of gradient in the acceleration.

A steeper gradient can be created by making the centrifuge smaller and spinning it at a higher rate. Thus, for example, it would be possible to have a space centrifuge with a normal, one-g, Earth-surface type of acceleration on its end or rim while several feet inward toward the hub, the acceleration could be half that amount.

Nobody has really given too much consideration to the fact that it is possible to obtain acceleration gradients in space for industrial purposes. It's not really possible to get an acceleration gradient from one-g down to zero-g here on Earth because of the Earth's gravitational field. So we haven't even developed or considered any industrial process possibilities that might make profitable use of this unique potential of the space environment. What can industrial engineers do with acceleration gradients made to order? They could, for example, create a combined fractional distillation unit with a separation process where solids of different densities are centrifuged out of the process at different points in the distillation scheme. It is very difficult to conceive of such future applications, however, because we are still attuned to thinking of processes only in a one-g field here on Earth. Once engineers get out into space and begin to live with these new conditions on a daily basis, we will begin to see some totally unexpected new products.

The centrifuge principle, of course, can be used as a separation process in space where there are no density-driven separation mechanisms. Here on Earth, if you want to separate milk and cream, all you really have do is let it sit for a time; the cream, being of less density, comes to the top where you can skim it off for coffee or butter or real ice cream. You can do the job faster with a centrifuge that creates a higher acceleration and a steeper acceleration gradient.

In space, milk and cream will not separate naturally in zero-g. You must centrifuge them to get them to separate.

In a like manner, if you have a liquid and a gas mixed together, the gas will remain in little bubbles throughout the liquid in zero-g. There were some problems along this line in both the *Apollo* and *Skylab* flights. In the *Apollo* flights, the crew's drinking water came from the hydrogen-oxygen fuel cells that also provided electricity. It was almost impossible to get all the tiny hydrogen bubbles out of this water. Several centrifuge-type measures were tried, but the centrifugal accelerations in spinning a water bag were a little too small to remove all the hydrogen bubbles. As a result, the astronauts drank gassy water. This caused a certain amount of crew flatulence and discomfort.

A small laboratory centrifuge would have solved the problem.

Two materials of different density will remain mixed in zero-g, as shown in illustration on page 112. These can be two liquids of different densities, a solid and a liquid, a solid and a gas, a liquid and a gas, or two solids of different densities. Normal separation processes caused by Earth's gravita-

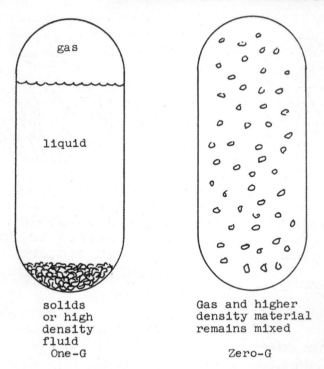

solids
or high
density
fluid
One-G

Gas and higher
density material
remains mixed

Zero-G

It is possible to obtain a uniform, homogenous mixture of materials of different densities in zero-g and maintain this uniformity.

tional field are shown on page 113. You can't get this sort of behavior in zero-g unless you apply acceleration by means of a centrifuge. You can obtain very good separation of materials or, by using an acceleration gradient in your centrifuge, you can obtain separation of only one component of a mixture while leaving the others mixed.

As a matter of fact, you don't even need a centrifuge to do this, because you can use the salient feature of the space environment, weightlessness, to cause a liquid to levitate in mid-air. After all, remember that in zero-g, liquids do not need a container; they are objects in their own right and will usually float freely while assuming a spherical or semispherical shape.

As shown on page 113, it will be possible to take a floating

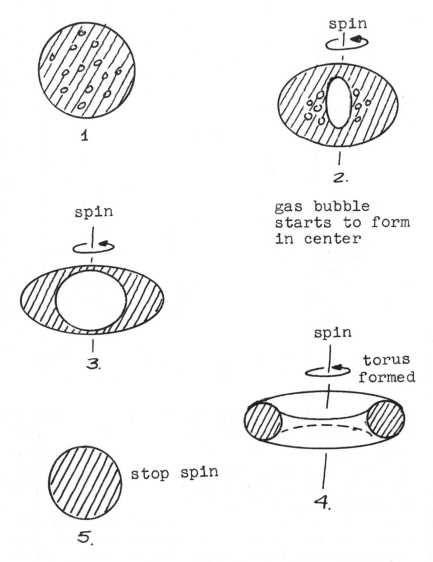

1

spin

2.

gas bubble
starts to form
in center

spin

3.

spin

torus
formed

4.

stop spin

5.

Degassing and density separation in zero-g by the spin-up technique.

spheroid of a mixture of liquid or liquid/gas materials and start it spinning; there are several ways to make it spin. As it spins faster and faster, it begins to flatten out into an ellipsoid, finally becoming a torus, or doughnut shape. The gas or lighter materials have by that time separated itself from the more dense material. The spin is then stopped, and the liquid material, now purified by separation, reassumes its spherical shape.

By using such a technique, it is possible to achieve an excellent separation of materials, achieving the highest purity in the process and eliminating any possible contamination from a container.

With this sort of process, space engineers will be able, for example, to produce vaccines of very high purity, solutions that are not contaminated, pure metals, and other uncontaminated chemical compounds. In many cases, we don't know the characteristics of pure compounds because it is not possible to produce them here on the Earth's surface. We will certainly see more of them, however, as the Third Industrial Revolution progresses.

SPACE INDUSTRIAL PROCESSES
(Adapted from Wuenscher)

1. FREE AND CAPTIVE SUSPENSION.
   a. Crucible Support, Wetting and Nonwetting.
   b. Sting Support, Wetting and Nonwetting.
   c. Electromatic Field Support.
   d. Electrostatic Field Support.
2. MIXING.
   a. Mechanical.
   b. Induction.
3. SEPARATION/PURIFICATION.
   a. Centrifugal Separation, Free or Container.
   b. Velocity Separation, Condensation or Selective Membrane.
   c. Electrophoresis.
   d. Magnetic Separation (mass spectrometer).

    e. High-Vacuum Refinement, Centrifugal or Marangoni

4. ALLOWING + SUPERSATURATION.
    a. Premixed Powder Melting.
    b. Thermosetting or Diffusion Alloying.

5. CASTING.
    a. Surface Tension Casting and Free Casting.
    b. Supersaturated Alloy Casting.
    c. Composite Casting, 2-State or 3-State.
    d. Adhesion or Layer Casting.

6. LIQUID STATE FORMING.
    a. Blowing.
    b. Electrostatic Field Forming.
    c. Composite Casting, 2-State or 3-State.

7. CONTROLLED DENSITY PROCESSING.
    a. Dispersion Foaming.
    b. Vaporization Foaming.
    c. Variable Density Casting.

8. DEPOSITION.
    a. Adhesion Coating.
    b. Galvanic Plating and Coating.
    c. Vapor Deposition.

9. SOLIDIFICATION.
    a. Amorphous Solidification.
    b. Controlled Crystallization.
    c. Single Crystal Solidification.
    d. Supercooled Coining.
    e. Zone Refining.

10. MELTING.
    a. Complete Melting, Low and High Viscosity, Overheated.
    b. Partial Melting, Matrix Melting in Cermets.

11. VAPORIZATION.
    a. Fractional Distillation.
    b. Pressure-Drop Vaporization.
    c. Freeze-Drying.

12. NUCLEAR PROCESSING.
    a. Fission Breeding.

   b.   Fusion Breeding.
   c.   Irradiation.
13. CHEMICAL PROCESSING.
   a.   Polymerization.
   b.   Free Radical Chemistry.
   c.   Free Atom Chemistry.
14. FERMENTATION.

CHAPTER

9

And More of the Same

THOUSANDS OF YEARS ago, some unknown, unsung, and unremembered genius either deliberately or accidentally—probably the latter—mixed together copper and tin in a molten state. When this mixture cooled and solidified, it was harder, stronger, and more durable than either of its two constituents. It was called bronze. It was the first of the materials we call alloys.

Down through the years, alloy technology has progressed with increasing rapidity. Bronze did not corrode, but iron did rust. This probably led some superconservative types to throw away their iron weapons and go back to bronze. But not for long. Eventually, the science of metallurgy built its foundations and expertise, finally creating stainless steels that do not rust. There are only two of the thousands of different alloys that now exist. It is now possible, to some extent, for a metallurgist to tailor the properties of an alloy to meet a given set of specifications—in some instances. Metallurgy is still almost an arcane art, often very empirical in nature.

In spite of rapid progress in recent times, all alloying is still basically the same process that was used in the creation of bronze: Mix two or more molten metals together in varying proportions to achieve as close as possible a true homogenous mixture. Then let this mixture cool and solidify.

Sometimes this doesn't work very well because of impurities in one or more of the alloying metals or impurities picked up from the crucibles. Some alloys require working after formation to remove impurities and to improve various aspects of performance.

Sometimes the basic, ancient alloying process doesn't work very well because of density differences between the metals. One of them may have a tendency to rise to the top of the mix before it cools and solidifies. This does not produce a very uniform alloy; its properties in one part of the mix may vary wildly from those in another part. Density differences may thus prevent the alloying of certain metals into mixtures that might have some very desirable characteristics.

Other metals may not mix well or uniformly because of wide difference in their surface tensions. They don't wet each other. They won't mix. By analogy, they are like oil and water. Italian salad dressing behaves this way; you may be able to shake it up to get the oil and vinegar dispersed in each other, but they very quickly separate into their components.

There are a lot of new and fascinating alloys being produced on Earth at this time. For example, gallium arsenide is a semiconductor that is used in modern electronics for such solid state devices as the light-emitting diodes in your electronic slide rule or desk calculator readout. It's also used to make Gunn Effect avalanche diodes for microwave radio sources and, as such, is an important material for long-range microwave radio links and radar. It's used in a number of highly specialized electronic semiconductor devices such as limited space-charge accumulation devices.

If you look at the Periodic Table of the Elements, you'll see that bismuth is of the same period as arsenic. Logic tells us that we should be able to obtain some very interesting new semiconductor devices if we could alloy gallium and bismuth.

But we can't. Bismuth won't mix with gallium. It is an immiscible combination here on Earth.

However, experts now believe that it may be possible to achieve a gallium bismide alloy in zero-g, where density differences will not cause separation and where it may be possible to get a homogenous mixture of the two metals. They also believe that this alloy may exhibit some interesting and unique semiconductor properties, although they don't know what these might be yet—because they can't make gallium bismide on Earth!

There is a broad field of investigation open for new alloys formed under space conditions. Of all the metals in the Periodic Table, only 7 of them are primary alloying metals: iron, copper, zinc, lead, tin, aluminum, and magnesium. There are 15 other minor metals. Together, these are combined with 40 other elements for making alloys. Some of these metals don't combine here, and some don't combine with the 40 other elements. There are an astronomical number of possible alloys that could be made by homogenous alloying processes available in space because there are not only the "binary" alloys everyone thinks of that are formed by 2 metals, but also ternary alloys (3 metals), quaternary alloys (4 metals), and even quinary alloys (5 metals).

Some unique materials are going to emerge from space alloying technology. It's too early yet to say for certain exactly what these will be and what their properties may be. It is quite certain, however, that some of them will be "wonder metals" with what seem like near-magical characteristics to our twentieth-century minds. They may have fantastic ultra-high temperature characteristics. They may possess the ability to be room temperature superconductors of electricity. They may exhibit unusual semiconductor properties.

The industrial environment of space is also going to make possible the first significant departure in alloy technology since the beginning of the Bronze Age. Since that time, nearly all alloys have been formed in exactly the same way: Heat both metals to liquid form, mix them together, and let them cool. They can then be cold-worked, hot-worked, heat-treated, or otherwise processed. But they are made in the hot and molten stage. There is an important reason for this: Most metals won't mix except at temperatures well above their melting points.

However, for years dentists and jewelers have used an alloy that is technically called an amalgam, an intermetallic compound that is usually a mixture of solid silver and liquid mercury. An amalgam is an alloy formed of a solid component and a liquid component, a metal of high melting point coupled with a metal of low melting point. The resulting mixture has characteristics that are different from either component and a melting point that is higher than that of the low-melting component.

The biggest problem with fabricating a silver-mercury amalgam is getting them to mix homogenously. A dentist makes small amount of amalgam to fill cavities quite easily by vigorous shaking or by direct mixing. The amalgam sets or cures at room temperature as the crystalline structure forms. It is very difficult to make large, industrial quantities of such amalgams.

On Earth, that is. Not so in space.

It is also conceivable to form true homogenous alloys consisting of a low-melting-point metal and a high-melting-point metal and then "curing" them at a higher temperature. The big problem with these "thermosetting" alloys on Earth is the fact that they are extremely sensitive to mixture distribution homogeneity; you just can't get a good, thorough mix with the same characteristics all the way through the alloy. Because of density difference, the two metallic components tend to separate. But in the density-insensitive realm of zero-g, we may be able to get them to combine without concern that they will separate and cause variations in the mix. Thus, we can make large quantities of them.

When we are able to make large quantities of these amalgams and thermosetting alloys in zero-g, we may be able to combine them to form some very interesting composite materials by using the superwhiskers that we will be able to make in space.

Metallic whiskers are still a new technology. A whisker is a very thin (less than 0.005 inches) and long (a quarter of an inch or longer) single metallic crystal that possesses exceptionally high tensile strength, many times that attributed to the metallic substance in the engineering handbooks. They are combined with other materials of lower strength to form a composite material. Fiberglass is an example of a low-temperature, low-strength composite material. Superstrong metallic whiskers have been combined with such low-strength materials as carbon (graphite) to produce exceptionally strong materials.

But there is a problem with whiskers: They cannot be grown to long lengths in a one-g field because they break off as they are being made. With any luck at all, it may be possible on Earth to form a whisker an inch long.

Whiskers are now commonly made on Earth by growing them around a thin tungsten wire so that they can be made long enough to do some good in a composite material. This creates enormous complications in the metallurgy of whiskers. Naturally, the tungsten wire interacts with the whisker material to some degree. Such tungsten-wire-core whiskers have lower-than-theoretical strength. And because one can make a tungsten wire only so small without having it break off as well, this technology limits how small a whisker can be made.

But in the zero-g of space, where there is no gravitational force to create dislocations of the metallic crystals in a whisker and cause them to break off, we will be able to form whiskers without the tungsten wire center. We will be able to make whiskers very small. And we will be able to make them very long. We will be able to create whiskers that are really microscopic, single-crystal wires with tensile strengths of over a million pounds. We haven't even started to guess the applications of such a material yet!

These superwhiskers can be combined with a host of other low-strength materials to create ultra-light materials of incredible strengths, strong materials with outstanding chemical characteristics, and strong materials with unusual elastic properties.

Furthermore, these supercomposites can also be made in space under zero-g conditions to insure absolute uniformity and homogeneity. The whiskers won't float in their base material, and it will be possible to form composite materials out of whiskers and bases that normally will not mix with or wet each other.

It is not necessary that a composite material be solid either. The base material can be a foam or cellular material.

Foams and cellular materials are rapidly becoming part of our industrial and commercial product lines. Most of them are made from organic or plastic materials. The range of foamed materials runs from inexpensive polystyrene foam with microscopic pore sizes to polyurethane sponges with large pore sizes. Foamed plastics are widely used because of their excellent heat insulation properties as well as their light weight. However, in common with all foamed materials, their strength degrades as the percentage of their volume devoted to pores increases.

It is possible to obtain foam metallic materials of aluminum or steel. However, these foamed metals are expensive because they are extremely difficult to make with uniform and repeatable characteristics in a one-g gravity field. Again, this is due to density differences between the melted, liquid metal and the gas that is injected into it to produce foam voids or bubbles. The metal has to be cooled rapidly to prevent the bubbles from rising to the surface. With a great deal of care, it is possible to obtain reasonably uniform pore distribution through a foamed metal batch, but it is difficult. Hence, it is expensive.

However, if a foamed metal is made in zero-g in space, it will have uniform distribution of the pores throughout the batch because the bubbles of gas will not rise owing to density differences. This will mean that the cooling of the metal will not have to be hastened, and this will permit metallic foams with better crystalline structures in the metal.

Today's Earth-formed foamed metals are composites of metal and air. But there is nothing to keep them from being mixtures of metal and other gases. In the zero-g of a space factory where the atmosphere surrounding the manufacture of foamed metals can be controlled to suit the product, some very unusual and unique foamed metals are going to be made. For example, there are places today where we might use foamed metals instead of foamed plastics, thereby permitting the particular device to be operated at a higher temperature.

In spite of the fact that we will be able to make better, more uniform foamed materials in space, they will still suffer from the fact that they will not be as strong as solid materials of the same type. We will be able to correct this, however, by making a metallic-whisker-reinforced foam material. The addition of superstrength long space-made whiskers to foamed materials will give them increased strength. Thus, we will be able to make in space, and only in space, fiber or whisker-reinforced foamed or cellular materials that are extremely light and very strong.

We will even be able to beat Mother Nature at her own game, because she originally developed or evolved the fiber-reinforced organic cellular material called "wood" that can be cut, shaped, sawed, drilled, machined, and fastened just like a plastic. "Space wood" made from a wide variety of foamed materials and strengthened with long fibers or whiskers uniformly deposited through the material can be vastly superior to wood for a number of applications, including dwelling construction. Such a material belongs to a group of materials that are, even today, called engineered materials. One picks the characteristics he wants and those he does not want, and then designs and engineers a material having those specifications.

There is another type of material that we are going to be able to make in space that we cannot make in large quantities on Earth or, perhaps, cannot make at all in some instances. This is the class of material characterized by the large-area ultrathin film.

In the zero-g environment of space, we will be able to make very large, very thin membranes by means of surface tension drawing with no support. On Earth, the production of ultrathin membranes requires the use of base or substrate to support the membrane against gravity forces. This limits the thinness to which a membrane can be drawn because, after you have formed the membrane on the substrate, you have to strip it off the substrate.

And if the membrane is one or two molecules thick, it is very fragile and tears easily.

Obviously, we will not have to support a membrane with a

substrate in zero-g. We can make it very, very big without worrying about the gravity field pulling on it and distorting or breaking it. It is rather like blowing a gigantic soap bubble. On Earth, gravity forces eventually cause the soap bubble to break. In space, you would be able to blow a soap bubble of theoretically infinite size!

One method of forming a large, ultrathin membrane in zero-g is shown on page 125. Here, a large bubble is blown first. The bubble material can be any liquid or molten material you wish to form into a thin film. The large bubble is then deposited on two frames as shown, and the bubble remaining between the frames is broken. Two large, ultrathin membranes remain on the frames. This is but one of several methods being discussed and investigated for forming large thin membranes in space.

Theoretically, it would be possible to form a membrane the size of a football field or larger with a thickness of a single molecule.

What would we do with such a large, ultrathin membrane?

What did engineers do with the superthin films they were able to make on Earth? The entire field of integrated circuits grew from such work. Minicomputers and electronic pocket slide rules came from this work. The potential of supersize ultrathin films made in space is probably greatest in the electronic and computer areas, but we may also see applications arise in photography, filtration, refining, pharmaceuticals, and solar energy technology. Because of the absence of gravity in space, it is also going to be possible to produce some very, very pure substances. Not only will we be able to use some of the refining and separation techniques discussed earlier to eliminate unwanted substances, but we will also be able to handle materials in weightlessness without permitting them to touch anything.

In zero-g, of course, everything floats. We can therefore "levitate" our pure materials in whatever pure gas atmosphere we wish or in a pretty hard vacuum if we wish. The superpure materials can be formed free of container contamination. These materials can be moved by electromagnetic fields, if they are metallic and conductive, or by jets of non-

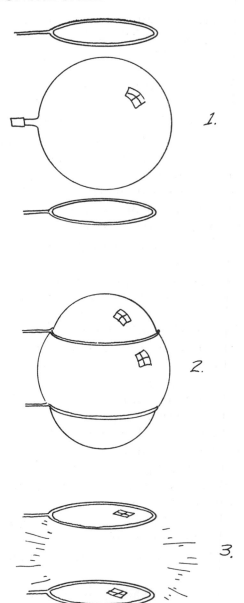

Production of large, thin, flat membranes by the bubble and frame method in zero-g.

Electromagnetic "levitation" and heating of metals in zero-g can be accomplished by the application of radio frequency energy to coils around the sample. Heating takes place by induction; positioning and handling is accomplished by varying the r-f energy in various coils.

contaminating gas, or by surface-tension phenomena such as Marangoni Flow, or by clever engineering of centrifugelike devices.

And it will become quite practical to make these superpure materials in very large quantities in space.

High-purity materials are usually very expensive, mere laboratory curiosities, until new processes are discovered and perfected to make them available in quanties. Aluminum is a good historical example of this.

We now take aluminum for granted. Housewives use aluminum foil to wrap garbage, a complete throwaway use that would have staggered the imagination of our ancestors. In 1827, aluminum was one of the rarest of all metals, even though it is one of the most abundant elements in the Earth's crust. Because of the very great difficulty in extracting it from its ores, it sold for $160 per pound in 1827 dollars, a small fortune in those days. Its utility was confined to laboratory experimentation by wealthy scientists and chemists.

In 1886 Charles M. Hall developed the electrolytic process for the extraction of aluminum from bauxite ore using a cryolite catalyst. Within a few years, the price of aluminum dropped to 15 cents per pound.

And production soared from a few pounds per year to millions of pounds per year.

And eventually made possible the modern airplane.

And the pop-top beer can.

We are likely to see similar things happen as space industry picks up steam and uses the unique characteristics of the space environment to produce things such as ultrapure metals.

Superpure metals often have physical and chemical properties that are quite different from their slightly impure forms. Beryllium is an example of this that we already know about. Here on Earth, it is usually produced by an electrolytic process from the double fluoride K_2BeF_4. It's not pure, and it is brittle. It is also hard enough to scratch glass. This makes beryllium very difficult to work, to machine, to be shaped and cut and formed using our industrial machinery.

Since beryllium is the only stable light metal with a high melting point, it is used in gyroscopes, accelerometers, and other precision parts of the inertial guidance systems now widely used in commercial aircraft and ocean vessels. Because of its excellent heat conductivity, it is used in heavy-duty brake drums. Beryllium is used as a moderator and reflector in compact, high-power nuclear reactors. And when alloyed with aluminum and iron, it forms some very strong, very ductile metals. Beryllium copper is the material used to make tools for nonsparking applications.

What else can we do with beryllium when it becomes available by the ton at low prices at the local spaceport? What can be done with inexpensive, easily obtained, ductile beryllium?

It's like asking an 1830 engineer what he could do with aluminum and if he could make an airplane out of it. Both concepts were totally impossible, impractical, harebrained ideas at that time.

So do not automatically assume that some of today's very exotic materials and techniques that are laboratory experiments will remain that way once engineers have the ability to process those materials in space with new technology that works *only* in space.

CHAPTER

10

Engineering with Atoms

THE TOTAL environment of space is still so new to our experience that we have to think hard to discover some of the things we are going to be able to do out there. But the results of these mental labors should not be considered as mere pipe dreams. Engineers have to do this sort of thing all day long every day of their careers. Without even seeing the ocean, they have to be able to design a supertanker or offshore platform to withstand the environment and do its job correctly and with economy. They do not have to be pilots to design and build airplanes. With numbers and training and a knowledge of the basic way the universe seems to work, they can make something work on paper before it ever gets into the shop.

Size has nothing to do with it, either. The device or system can be as large as the Golden Gate Bridge or a *Saturn 5* moon rocket. Or as small as a single electron moving in an integrated circuit chip or a computer's bubble memory.

This is sheer magic to most people who don't understand it and follows a paraphrase of Arthur C. Clarke's Second Law: "Magic is a technology that you can't understand."

So when we talk about these things that we will be able to do in space industry, remember that most of them are based on material that is already published in *Mark's Mechanical Engineers' Handbook* by Theodore Baumeister or the *Handbook of Chemistry and Physics,* edited by C. D. Hodgman, or any one of a number of other common scientific handbooks that are on nearly all engineers' bookshelves. Plus some of the latest data to come out of earthbound research labs.

For example, let's just consider two of the space environmental characteristics that we've discussed but have not yet delved into: low temperatures and high vacuums.

These two characteristics alone are going to bring to rapid maturity an entirely new area of basic chemistry that is *really* new!

During 1971 Philip Skell and J. J. Havel of Pennsylvania State University announced the first results of their preliminary work in the new field of free atom chemistry.

Free atom chemistry is a relatively new area of technical investigation primarily because of the lack of suitable techniques for experimentation. It is not easy to get a supply of free atoms of a single element, and it has been even more difficult to get them to react with other free atoms. The main source of free atoms for experimental work—which is the only work that has gone on to date—is boiling them off a hot wire in a vacuum. This means that the temperatures of these free atoms are very high indeed, which in turn means that they are moving quite rapidly.

These hot-wire free atoms may have temperatures of several thousand degrees Centigrade or more. It is not possible to get them to react with compounds and organic molecules that decompose at temperatures of a few hundred degrees Centigrade. These compounds come apart at high temperatures, and there are no controls on the experiment to let you know what happened to which atom.

Skell and Havel used earthbound, brute-force methods in their original experiments. They got their free atoms in the classical way by boiling them off electrically heated wires of the material whose atoms they wanted. They let these free atoms proceed through a very hard vacuum and let them impact upon target substances that had been supercooled with liquid hydrogen.

Thus, they were able to study reactions of free atoms with low-temperature molecules because, even when the superhot free atom hit a target molecule, it did not raise the target molecule temperature from the supercooled state to the decomposition point. This is rather like a macroscopic analogy of dropping a glob of hot solder onto a block of ice; it melts the ice, but does not turn it into steam. The glob will melt its way into the block of ice, and the melted water will then freeze to produce a shiny glob of solder imprisoned in the block of ice.

The work of Skell and Havel in free atom chemistry is important because we have not generally been able to study the discrete interactions between a single atom and another molecule. The province of chemistry has been that of studying the behavior of molecules, but this does not really tell you very much about the behavior, characteristics, or properties of the atoms that make up the molecule. Common table salt is a glaring example of this. The sodium atom burns fiercely in the oxygen of the atmosphere at room temperature. The chlorine atom is highly toxic to organic life. Yet, together in the salt molecule, a totally different set of characteristics exists. In fact, life could not exist without salt, in spite of the fact that the two atoms that make up the salt molecule are, by themselves, lethal.

In a similar fashion, the behavior of a single hydrogen atom is quite different from that of the biatomic hydrogen molecule in all its different metastable spin states.

Trying to determine the chemical properties of atoms by studying their reactions in molecular form is like trying to study an individual by investigating the reactions of a crowd of people.

Thus, there have been some fascinating results to date from the initial free atom chemical experiments of Skell and Havel.

For example, they have studied some of the reactions of free platinum atoms with various molecules.

Platinum is a "noble" metal with a high resistance to corrosion. It appears to be an "introverted" atom that doesn't like to get together with anything else. Normally, it will react only with the strongest oxidizing agents such as aqua regia. For this reason, it is in sharp demand for jewelry because it will not leave a green stain on your wrist or finger. This also makes it very hard to refine or reduce from its ores because when it does form compounds, it does so with lasting friendship, so to speak. And it does not like to be pried apart from any molecule it has happened to react with.

This also makes it very difficult to study.

But by using the free atom chemistry techniques they developed, Skell and Havel found that free atoms of platinum were highly reactive with some substances. They obtained free platinum atoms by heating a platinum wire to incandescence with electricity in a vacuum. They permitted the platinum atoms to react with hydrogen-cooled targets in a vacuum.

Propylene is a very innocuous organic molecule with a very low decomposition temperature and a very limited list of elements with which it will interact. As polypropylene, it is a long-chain organic molecule that is a common plastic these days.

Skell and Havel got free platinum atoms to react with propylene molecules.

And they have worked with more than 40 metals as free atoms, getting them to react with other molecules in very unsuspected ways. Reactions and interactions took place on the supercooled targets that had never been seen before.

Free atom chemistry is only a few years old. There is a lot to learn. Earthbound experimentation is difficult and expensive. Materials are made in microscopic, laboratory quantities at great price.

But free atom chemistry holds forth the promise of entirely new classes of organometallic compounds with completely new and unsuspected properties. But rapid progress in this area and industrial production of some of the useful compounds that are discovered depends upon the ready and cheap availability of two things: very high vacuum and very low temperatures.

Space has these two characteristics. There is a lot of vacuum out there. And temperatures approaching absolute zero can be obtained by relatively simple isolation of the process from radiant heat sources. Put the process on the shady side of a space factory and equip it with suitable heat radiators to eliminate any heat generated in the process . . . and the process temperature drops to that of interstellar space, a few degrees above absolute zero.

Here again is a new process that can be done in space at lower cost and complexity. And that can produce yet unknown materials that we can't even carry out a market study on yet. A speculative venture into technology that depends upon serendipity for a payoff. But we will not know what can be discovered until we try. If we do not try, somebody else may. And who is willing to bet that the results will not affect his current earthbound, money-making products?

For the same reasons—very expensive earthbound facilities and lack of useful techniques—free radical chemistry has also suffered from a lack of progress since the first free radical was discovered in 1900 at the University of Michigan by Gomberg, who succeeded in isolating the methyl radical.

A radical is a fragment of a molecule that is ionlike but does not carry an electric charge as an ion does. It does have a chemical valence charge. An example is the ammonium radical NH_3^+ of ammonium nitrate or the sulfate SO_4^{-2} radical from sulfuric acid.

Radicals are highly energetic, but have lifetimes measured in milliseconds. They are very "gregarious"; they tend to combine with other atoms and molecules at the slightest provocation, depending on the radical and the chemical circumstances.

As with most chemical activity, extremely low temperatures tend to increase the lifetimes of free radicals. Therefore, most of the research in the field that has been done thus far has been carried out in supercold cryogenic fluids.

The field of free radical chemistry and energy generation was surveyed thoroughly by the U.S. Air Force Office of Scientific Research in the late 1950s with the conclusion that the technology was not yet to the point where significant breakthroughs of interest to the Air Force were possible.

The Air Force was looking for a vastly improved rocket propellant.

The most exciting possibility in free radical chemistry was and still is monatomic hydrogen.

Hydrogen normally exists in the diatomic state H_2 and in various forms, depending upon the spin relationship between the two hydrogen atoms.

If a method could be found to produce and stabilize monatomic hydrogen—single-H—and permit it to recombine with itself to form the more stable diatomic hydrogen, it would release energy equivalent to 90,000 BTUs per pound. This is roughly three times the energy available by burning diatomic hydrogen in diatomic oxygen.

Atomic hydrogen—single-H—would not only be an outstanding rocket propellant because of its high energy and low molecular weight, it would also rival nuclear fission as an energy source right here on Earth.

Solutions with as much as 1 percent single-H have been formed, but it would require mixtures of at least 10 percent single-H to begin to be effective as an energy source. Four major technical breakthroughs need to be made for single-H to become useful: (1) we must learn how to generate large amounts of single-H; (2) we must find out how much of it can be concentrated in a stabilized state; (3) we must discover how long we can stabilize a maximum concentration; and (4) we must work out a way to recombine single-H in a controlled fashion and in a useful manner.

It rather appears that we are at the stage of the game with free radical chemistry of knowing that it can be done theoretically, yet not knowing how to do it practically. It is very much similar to the state of affairs in nuclear energy in the early 1930s.

All that is needed is the equivalent of the Hahn-Strassmann experiment of December, 1938.

In nuclear research, fission was a theoretical thing until

Otto Hahn and Fritz Strassmann performed the critical experiment that confirmed it and showed the way to go. The Hahn-Strassmann Point in any technology, according to technological forecasters, is that critical point where an experiment or finding suddenly makes a whole new area of technology immediately feasible. Before the Hahn-Strassmann Point, everything is impossible and nobody can get anything to work. After the Hahn-Strassmann Point has been reached, the technology takes off and rapid development follows.

The very low temperatures and the controlled matter environments easily obtained in space will permit free radical chemistry to take on a new lease on life, perhaps to reach its own Hahn-Strassmann Point. This could occur as early as the first part of the 1980s when the *Spacelab* becomes available, lofted by the Space Shuttle.

When it does happen—and it will because of the increasing need for new energy resources—it will not only provide us with a new energy source for earthbound use, but also an important key to really cheap space transportation.

Single-H, used as a rocket propellant, would be as good as or better than nuclear rocket propulsion of the NERVA type. Furthermore, single-H as an energy source is definitely nonpolluting and is not radioactive.

It would create as radical a change in space transportation as the jet engine created in air transportation.

Engineering with free atoms and free radicals not only will permit us to put chemical elements together in a gross fashion as we have been doing on Earth for centuries, it also means that we will be able to do true molecular engineering by putting atoms and molecules together in certain definite patterns.

Nature does this, and the result is a crystal.

Industrial crystals have been grown for years, and crystals are an important part of our industrial scene, all the way from quartz crystals for electronics to the metallurgy of the steel industry.

The properties of crystals fascinate people; carbon crystals—otherwise called diamonds—are highly prized by women and industrialists alike. Without crystals of various substances, modern electronics would not exist; in fact, without crystals, electronics and its allied telecommunications technologies would never have gotten started. We have a great deal to learn about crystals and the properties of crystalline substances. Dr. Henri-Marie Coanda once pointed out to me that all electronics and computers can be considered to be inorganic crystals while we can be thought of as organic crystals.

Since crystals are characterized by the orderly array of their constituent atoms, a flaw in a crystal occurs in several ways when this orderly array is disturbed. It is very difficult—impossible in most instances—to make a perfect crystalline structure here on Earth except in very small sizes. Discontinuities are created, dislocations of crystalline structure occur, and crystals contain imperfections because of our ever-present gravity field.

Some of the first experiments in space processing aboard *Apollo* and *Skylab* involved the growth of crystals in zero-g.

We know now that in space we can grow perfect crystals of very large size.

This fact will have immediate impact upon several industries in the early 1980s—electronics, communications, bearing technology, electric power, turbine engines, computers, and glass, to name but a few.

This is one crystal ball that is not cloudy.

Space, a place most devoid of matter, is going to be a place where we can put matter together piece by piece in ways that will be useful to us. What we do not find in space in the way of raw materials, we may be able to make.

On the night that the human race first set foot on another planet, July 20, 1969, a profound statement was made on CBS network TV by an engineer, writer, and retired naval officer named Robert A. Heinlein: "I'm not at all sure just how things are going to be done. I'm simply certain that they are going to be done in a big way. . . . There's just one equation that everybody knows: $E=mc^2$. It proves the potentiality whereby man can live anywhere where there is mass. He doesn't have to have any other requirement but mass, with the technology that we now have. . . ."

CHAPTER
11

The Space Factory

WHEN ONE LOOKS at a factory here on Earth, one sees certain things. There are warehouses, storage areas, junkpiles, workers' transportation vehicles, offices, and one or more buildings in which the industrial operation actually takes place. In almost all cases, the building encloses the industrial process or operation and provides, inside its walls, the environment needed for the processing. This factory building also encloses people and provides a comfortable year-round environment for them . . . usually.

The forthcoming factories in space will be different in some aspects and the same in others.

The first space factories are going to be transient things such as *Spacelab* and its variations that ride into space in the cargo hold of a shuttle and later are returned to the ground. Their historical analogy is the whaling ship from Gloucester or Mystic, a multipurpose vehicle that not only collected the raw material but also processed it right aboard the ship. The workers lived aboard the whalers. They stayed out until they could process no more products or until they ran out of food and water. Modern whaling ships still do the same things, but it is less likely that they will run out of food and water. Life aboard or in one of these early transient space factories can be glimpsed today by studying the life aboard whaling and fishing ships.

These early up-and-down space factories will be very small in comparison with what will come within decades. And only small amounts of limited products will be processed in them. These will include vaccines, crystals, composite materials and items made from these composites, thin films, and special, highly engineered materials.

Eventually, within a few years, the time will come when it will be more economical to keep the process—whatever it happens to be—running continuously because of cost factors or demand. That will be the day when a space shuttle attains orbit, opens its cargo doors, and gently deposits in space a complete manned module with only a few technicians to run things. The shuttle will then return to the ground, and the first true space factory will be in orbit. In subsequent flights, the shuttle will bring up replacement technicians, consumables for the life-support system, and new raw materials for processing. The shift at the space factory will change, and the shuttle will go home with its valuable load of space products.

Soon another module will be deposited in orbit alongside the first one by a shuttle. It may be a duplicate of the first, put into space to expand the plant to meet demand. Shortly, there will be ropes running between these two space factories with spacesuited technicians making EVAs (extra-vehicular activities) between them. Perhaps the same number of technicians will be running both facilities. In due course of time, the two will probably be joined directly.

Through the years, this first space factory will grow like Topsy, just like any other industrial plant on the planet below.

Like its predecessors on Earth, the space factory will provide an enclosure to protect its workers from the environment. But there the similarity may end because the space factory probably will *not* also provide the environment for the industrial process. Space itself will be the industrial environment.

The first space factories will have to use the space environment because the unique characteristics of this environment, as we have seen, permit some unusual industrial operations.

And that is the only thing that will justify such Earth-orbital operations as the capital base of space industry builds upon itself.

Eventually, however, industrial research operations in the space environment will come up with new and better ways of doing things in zero-g, hard vacuum, and wide temperature extremes.

And the increasing pace of space industrial operations will require additional energy converters.

There will come a time very early in the Third Industrial Revolution when it will no longer be economical to lift raw materials and construction materials and space factory modules from the Earth's surface into orbit. The demand for space products will grow. The growing scarcity of Earth-resource raw materials will increase their price, and the transportation costs of the increasing amounts of raw materials from the surface to orbit will demand that new, cheaper sources be found. Expanding population will demand more products. And, eventually, antipollution laws and increasing energy costs will sharply limit industrial production on Earth or make it more costly than producing the same products in space.

People will then again go to the Moon.

They will go this time to stay. They will get there from one of the space factories in near-earth orbit, using either the currently planned Space Tug or something similar to Cole's Space DC-3. It's a toss-up at this time in terms of forecasting whether this will be a government job or an industry job. Perhaps a bit of both . . . because government is going to have to go into space to stay before the end of this century, for reasons that will become obvious and be discussed below.

These new landings on Luna will be for one purpose: to find raw materials. This means that these lunar exploration missions will become long-term missions. One cannot do a decent job of really detailed prospecting in a day or even a month; one must crawl around the landscape, picking up rocks, nosing into interesting corners of the selenography, and engaging in digging. Not all the raw materials on Earth are obtained by strip mining; the same is quite likely to hold true on Luna.

There may be trips to Mars, but it is very possible that Mars landings will not be made until *after* exploration trips to interesting parts of the planetoid belt. It is easier in terms of energy—both propulsive and life-support—to get into the planetoid belt than it is to land on Mars.

By this time—and the time-frame is probably in the late 1980s and early 1990s—we will have solved the next major technical hurdle to extended manned travel in space. We will have perfected the closed ecological life-support system.

The lack of such a system is the major item—other than lack of unaudited and unlimited government funding—preventing extended lunar missions or Mars landings with today's technology. Every space mission to date has been conducted with an open-loop life-support system requiring replenishment of consumables and overboard dump of wastes. The technology involved in closed systems has basically proven itself under laboratory conditions. We know how to recycle the breathing atmosphere to obtain a continuing supply of oxygen. We know how to recycle waste products and how to convert sewage into potable water again. We know how to provide a continuing food supply. We can do it in miniature, and we can do it in a laboratory. But we haven't done it yet on a large scale for dozens of people over a period of two to three years.

"Necessity is the mother of invention."

"When it's steamboat time, you steam."

When the need is really there for a closed ecological life-support system for long-term use in orbital space factories that are making money because of the products that can be produced in them, closed ecological life-support systems will be built, tested, qualified, and used in space. This is almost a sure-bet forecast for A.D. 2000.

Again, some perspective is required, as it usually is when forecasting the initial establishment of space industry, including lunar bases for prospecting and manned voyages to the planetoid belt in a period of thirty-five years.

At the time of this writing—1975—thirty-five years in the past is the year 1940. We are now halfway between 1940 and 2000. Does anyone really think that there will be less change between now and 2000 than there has been between 1940 and today?

By conservative estimate, over 75 percent of the products now available in the stores that are factory-made were not available in the year 1940.

By the year 2000 we should be well along toward mining the Moon and mining the planetoid belt. It is an integral part of the Third Industrial Revolution, and one that is amenable to logical forecasting because it is part of the logical capital development of space industry, which in turn is based on the hard, cold, logical forecasting of the business world. By this I mean the same sort of forecasting that goes into the planning of a bond issue and its eventual retirement.

Please also note well that this forecast of lunar and planetoid exploration isn't based on the assumption that a bunch of scientists will be going out there to poke around trying to learn the secrets of the universe for the sheer joy of it, although there will be some scientists doing just that. But they will be along for the ride. The forecast is based on the assumption that a bunch of explorers is going out there to poke around and dig stuff out and make something out of it and pocket some money as a result. That is one terrific incentive, as well as one hell of a risk, which means that the payoff will be tremendous for those who succeed first.

Within the first decade of the twenty-first century, one or more solar-powered lunar catapults will be built to lob raw materials into space for use by the zero-g space factories. These lunar raw materials will eventually replace those being lifted up from Earth for special zero-g, high-vacuum processing that can be carried out only in space.

Eventually, earthbound factories producing things from earth-source materials will close down because of obsolescence, stiff antipollution regulations, dwindling high-grade raw material sources on Earth, the high cost of industrial energy, and increasing labor costs on Earth. Many of them will be relocated, first in orbit, then on the Moon. This will be one of the greatest transfer operations in history because the facilities will not be transferred, only the operations. Eventually, many of the things made in earthbound factories will be made in orbital and lunar factories.

This transfer in the first half of the twenty-first century will first be made to Earth-orbital factories, then to lunar surface factories for a very simple reason. There are some processes that need gravity, and even the one-sixth lunar gravity is better than trying to produce pseudogravity in orbital factories.

If orbital space is an excellent location for industrial operations, so is the Moon. Everything that exists in orbital space also exists upon the Moon with the exception of the lunar gravity field.

To speak of polluting the Moon is patently ridiculous. The Moon has no biosphere to pollute. It is a natural and benign site for industrial operations. It may not only be a prime source for certain raw materials—we'll find out more about them in due course of time, after we've poked around a bit more up there—but it is also a stable foundation for factories and possesses a relatively low gravity field that makes exporting easy; the lunar catapults using solar energy can drop space-going boxcars to destinations on Earth as easily as this can be done from Earth orbit.

Although some lunar surface space factories may be on the lunar surface, it is quite likely that many of them will also be under the surface. The surface may be devoted to energy-collecting solar power screens.

According to Dr. Krafft Ehricke, it will be relatively easy to excavate sublunar caverns on the Moon for industrial use. We will be able to use nuclear explosives.

Please shed your earthbound prejudices! The use of nuclear explosives for excavation on the Moon doesn't affect the lunar ecology because there *is* no lunar ecology! It cannot pollute the Moon with radioactivity because there is nothing up there to pollute! The only danger is to those people who will eventually live and work in those excavations, and this is certainly measurable and controllable.

We may not erect space factories on Mars; we may not need to. We may not want to because we might disrupt the ecology there. We don't know if Mars has an ecology. We will get some hints in 1976 when *Viking* lands. We may simply want Mars' raw materials. We will certainly know more about Mars and its suitability for space industry within the next fifty years.

However, starting from very simple beginnings in 1981 when the first commercial space-processing operations begin in the cargo bays of the Space Shuttles, the space factory will evolve over the following fifty years until there are many of them in orbit around the Earth, many of them on the lunar surface, and even some of them in the planetoid belt.

Fifty years is within the lifetime of many of the readers of these words.

Orbital space factories will not look anything like the neat, well-organized, stark space stations of classic space art. They will be motley collections of parts hung together to enclose people, computers, and perhaps part of the industrial operations themselves.

If you knew about the individual development of an individual space factory, you would be able to locate the modules that formed its original operation. But, in common with every other factory and industrial operation anywhere, the space factory will grow over the years as the industrial processes it uses are modified and improved, as new capital equipment based on newer technology becomes available, as new by-products and modifications of the process are defined, and as engineers correct and fine-tune and improve efficiencies and lower costs just as they have been doing for years and years on Earth.

The typical orbital space factory in zero-g will be festooned with solar energy collector panels and, if the process requires it, hectares of solar reflectors to concentrate solar heat. It will carry huge, red, glowing heat radiators continuously shunting waste process heat off into space at the highest possible temperature that the radiator materials will withstand. Some of these are likely to be wired on and lashed up and hung on.

Skylab was only the first crude, primitive space lash-up.

The orbital factory will be surrounded with floating pieces of sky junk—parts of old equipment, parts of new equipment that haven't been installed yet, old modules acting as warehouses for finished products that must be protected from temperature extremes or radiation from the Sun, rocket-powered space scooters for transportation of people from factory to factory to orbital module, empty space boxcars waiting for their earthbound payloads, raw materials waiting to be processed, and probably the space equivalent of Mama Murphy's Greasy Spoon, Shop Lunches Put Up for Any Shift. . . .

By the end of the twentieth century, Earth is likely to be surrounded by several space factories in different orbits. Some of them may be in the same orbit and even within sight of one another in the same orbit. There is nothing that says that more than one satellite can't occupy the same orbit; witness the rings of Saturn. . . .

Space factories on the lunar surface and, eventually, on other planetary surfaces, will probably bear a closer resemblance to earthbound factories because they will have firm ground (not *terra firma*) under them. Again, they will sport hectares of solar energy panels and heat radiators.

Many surface factories may be underground for temperature control, solar radiation filtering, or ease of factory construction.

Lunar surface factories are likely to be located at or near a lunar electromagnetic catapult head so that their products can be more easily transported to market. However, economics may dictate that there are only a few catapults in the early years; after all, the catapults will probably be built using lunar or planetoid materials. The close earthbound analogy is the fact that every factory doesn't have its airport or wharf.

Lunar surface transportation from the factory site to the

catapult head could take many forms, and it is fascinating to consider what they might be and how they might appear if designed to operate on one-sixth of Earth's gravity.

Men have already driven wheeled vehicles on the lunar surface, and it is quite likely that we will see more wheeled vehicles there in the future as "moon trucks" carrying payloads and "moving on" from factory to catapult head and from landing stage to factory and from mine to factory, etc. They will probably be electric-powered as the *Apollo* moon buggies were. Early lunar "roads" may be nothing more than tracks across the regolith; eventually, they may be given a surface treatment so that heavier truck cargoes can be carried.

An interesting lunar transportation concept is the use of the funicular, or cable, car in slightly modified form. The one-sixth lunar gravity will allow trains of cars to be suspended from and travel along cables that are strung from towers. These "wireroads" may be the railroads of the Moon. Looking like earthbound electric transmission lines, they may be the quickest and cheapest surface transportation scheme that can be built on the Moon for fast transit. Like a railroad train or a monorail, they can operate in long strings of cars pulled by a single manned locomotive.

There has been some suggestion that lunar surface point-to-point transportation might be accomplished with small, suborbital rocket vehicles. However, this is less likely than a true surface transportation system because of the fact that a suborbital ballistic rocket system must expend energy to climb partway up a gravity "hill" to get where it's going, while a surface transportation system operates on a gravipotential plane. The only energy that will be required of a surface transportation system is that required to accelerate or slow down, plus overcome the friction inherent in the system; it won't have to worry about aerodynamic drag!

Space factories associated with small planetoids where planetoid mining may be going on will have even different appearances. Although eventually most planetoids that contain valuable raw materials will probably be moved to existing factory sites, there will be some planetoids that are so large—several miles or more in diameter—that it may be more economical to bring the space factory to them instead.

Such factory planetoids will look like rocks with strange forms of lichen hanging all over them.

The lichen will be human-built space factories, and the planetoids will be surrounded by swarms of busy space transportation ships.

Industrial man is messy by nature, and he will continue to be messy in space because the very nature of industry is to produce at the lowest possible cost because of competition, and even in the absence of competition, messy Industrial Man still exists. There is little difference in factory appearances on either side of the Iron Curtain that separates capitalistic free enterprise from socialistic state capitalism. In fact, if anything, free-enterprise capitalism has cleaner factories probably because of the importance of "image" in our culture.

In due course of time, once the Third Industrial Revolution is well established, sometime in the twenty-second century, a hue and cry will arise from groups of space "environmentalists" who want to return the Solar System to its pristine pre-Shuttle condition. They'll raise hell about the tailing piles on the Moon. They'll worry themselves sick that some alien civilization may be poisoned by the industrial wastes being lobbed into interstellar space.

But, by that time, space industry will be such an integral part of human culture and will be so far removed from the restored biosphere of Planet Earth that most people won't really care. The Earth will be a lovely place to live on again, and if we have to use up a few planetoids—lifeless at that—to achieve it, so be it.

CHAPTER

12

People and Human Institutions in Space

THERE IS ONE aspect about all space factories that will be vastly different from earthbound factories:

There will be no "front office" as we know it today.

The "front office" of a space factory will be a combined communications center and control center.

This is because the Second Industrial Revolution will have become quite mature by the end of this century. Automation technology is already at the point in 1975 where a great many industrial processes are carried out under automatic control with supervision from time to time by human beings. Automation has not yet peaked on Earth, primarily because of a need to continue to employ people in jobs that could be done by automation and partly because we have not yet attained the higher level of general education required of a work force in a fully automated industrial society.

But because it will be exceedingly expensive to provide continuing life support in space operations for the early space industry human operators, space industry is going to be built upon automation.

People will still be required, but they will be performing on a higher and more advanced working level.

We can perhaps better understand what the true function of people in space industry will be by considering that they are, basically, colloidal computers with an exceedingly versatile variety of functioning operational peripherals. The response time of their colloidal circuits is measured in milliseconds. They require an oxidizing atmosphere that is held within definite pressure limits and within certain temperature ranges for optimum operation.

There is an ancient (pre-*Sputnik*) adage of anonymous authorship that I ran into about 1950. It is an excellent rationale to justify the existence of a human being in an automated system: "Man is not as good as a little black box for certain specific things; he is flexible and much more reliable. He is easily maintained and can be manufactured by relatively unskilled labor."

(NOTE: When the word "man" is used herein, it is intended to be a contraction of the word "human" and does not connote any form of sexual chauvinism. Astronautics will not long remain the almost exclusive province of the male human. Arthur C. Clarke is absolutely correct in forecasting that the first human will be born on the Moon before the end of this century. I would like to expand that forecast to opine that the first human baby will be born *in space* before the end of this century. The Third Industrial Revolution is a *human* aspiration.)

Man is not preprogrammed. He can draw pragmatic conclusions from incomplete data. Although he can consciously handle only two variables simultaneously, and very slowly at that, he can unconsciously correlate an exceedingly large number of variables.

Mankind will take into space a recently developed symbiote: the crystalline computer. He has already done so. It is a classical and logical example of true symbiosis.

Crystalline computers have circuit response times that are measured in fractions of a millionth of a second. It is conceivable that, if computers were to become self-aware, they would become exceedingly bored trying to communicate with humans who are almost a thousand times slower. If a fast general-purpose computer could possess emotions, it would probably be exasperated with humans; to it, communicating would be like a human communicating to a very retarded child. Crystalline computers have very limited operational peripherals. We can provide them with hands, but nothing as elegant as our own wrist joint.

The human wrist joint is unique among mammals and is a supreme triumph of evolutionary engineering. A human can bend his hand at the wrist so that it is at right angles to his forearm, and he can still individually move every one of his fingers while rotating his hand nearly 360 degrees with his forearm. No other animal can do this. The mechanical equivalent of the human hand-wrist-forearm is a very complex machine.

Crystalline computers are also sensitive to radiation, to temperature extremes, and to atmospheric composition. They can operate in a vacuum, but they do not generally like low temperatures or high temperatures.

But the one basic difference between colloidal computers—us—and the crystalline computers is the rate of the evolution of the species. We are generating new models of computers almost daily as the "state of the art" progresses. But with man, we are stuck with Model Number One for a long time to come, at least until genetic engineering gets beyond the let's-make-a-monster stage and until Build-a-Man Kits become available for the kiddies.

Since there will probably be few, if any, unions to prevent automation of space industry, computers and automatic controls will be an integral part of the Third Industrial Revolution right from the very start. Computers will be used to exercise direct, fast control over complex and repetitive space industrial processes where information from sensors—temperature, pressure, light, etc.—can be fed directly to their data-processing circuitry. Some of this is already old hat even today, in the petrochemical industry, among others.

Although human millisecond response times may seem to be far too long when compared to the picosecond responses of a computer, this millisecond response time is often more than adequate, as we have found out over the eons. After all, most situations don't require any faster response times. A stopwatch doesn't make a very good calendar. So people will be required in the industrial operations of space, but not in the sense of thousands of workers per shift as in an automobile factory. Space industry will be much like today's automated petrochemical plants where a few highly trained people operate while others exercise the chain of decision making called administration and management.

People will be used to oversee and monitor what the computers control, to make on-the-spot decisions based on incomplete data when required, to change programs to fit sudden new circumstances, and to carry out the sort of work that was first done in space by the astronauts of *Skylab*.

People will not only be the supervisors and overlords of space industry, but also the builders and fixers . . . until we manage to come up with something equivalent to Rossum's Universal Robots, machines that are as flexible as human beings when it comes to that fascinating combination of a versatile brain in a versatile body.

People will build the space factories. Clad in pressure suits and floating free in space, secured perhaps by lines to shuttle ships, they will assemble and test and activate parts, assemblies, modules, and the other components of space factories. These will be pure and simple engineering jobs of the sort that go on today on the Earth's surface every time a bridge or building is erected.

Basically, for a long time yet to come, the people that go into space for industrial purposes will be specialists with years of training. As the living facilities in space grow to be less primitive and rudimentary, these specialists will not be transported to and from space living modules every few weeks, but will remain in space. In due course of time, they will have their families living in space with them.

We will know that we are really more than a planetary species once we take the basic reproductive unit, the family, into space to stay. It may turn out to be a different sort of family than we are used to thinking of here on Earth, but it will be recognizable as a family unit. The roles of the family members may be slightly different, and the children may be

schooled differently. But the transition will be made success-
fully.

We need only look back less than a century in our own
American culture to find analogies, particularly in the American West. But taking the family unit into space will be at
least an order of magnitude easier than making the family
unit survive in the West in the 1870s.

People will not only build the space factories and the ancil-
lary facilities as well as operate them, but they will also repair
and maintain them.

There isn't any automatic machine yet made that I have
ever seen that will operate totally without human care in the
form of maintenance and repair. Left to themselves, automatic machines destroy themselves. Have you ever watched a
teletype machine literally tear itself to pieces? Have you seen
an automatic gear lathe commit suicide? Even "long-life"
space satellites quit after a couple of years. And scientists
were ecstatic when the ALSEP experimental packages left on
the Moon continued to work for more than a year.

Other than a human being and other forms of organic life,
the only other self-repairing entity is a crystal, if it is placed
in a nutrient fluid.

But machines can be and have been designed and built
with failure-prediction and failure-reporting functions built
in. An example of a failure-prediction device is the oil pressure gauge that used to be standard dashboard equipment in
automobiles; by observing the level of oil pressure and its
fluctuations at idle speed as your motor grew older, you
could forecast when the rings and bearings would fail; you
could then take appropriate action to repair things before it
got too bad to fix. An example of a failure-reporting device is
the "idiot light" that replaced the more expensive oil pressure gauge on the dashboard of modern automobiles; the
light comes on only when an oil pressure switch on the motor
senses no oil pressure; it does not predict the failure of the
motor, but *reports* the failure as it happens, usually when it is
too late to do anything but shut everything off and hope.

A great deal of automatic equipment of a very complex nature will be used in space for industrial purposes. It will have both failure-prediction and failure-reporting devices of great reliability and sensitivity built into it. Much of this equipment will be computer-monitored. Many functions of an operation or device will be monitored; when an operation deteriorates to a predetermined level or when certain anomalies show up in the operation, this information will then be reported to a remote location, probably millions of miles away. For example, perhaps a high-temperature foundry is located in an orbit very close to the Sun, perhaps within the orbit of the planet Mercury. It may operate automatically because it may be very expensive to maintain proper life support for humans in that high-temperature environment.

When something starts to go wrong, when the failure-prediction circuits of the factory's computer sense an impending failure, the data is telemetered to a maintenance and repair center perhaps located in Earth orbit. There, a much larger computer reviews this data. If a predicted failure is indeed confirmed, the computer yells for help from a human being. A maintenance engineer reviews the data and decides to send a maintenance team to the factory. The team departs in a Space DC-3, arriving at the factory many weeks later. They are, of course, equipped with life-support systems that will sustain them in the high-temperature environment for a short period of time—several days or long enough to make the repair. Either they replace the malfunctioning component with a new one brought along for the purpose or they repair the one that is already there. When their job is done, they scat for home.

It may also be possible and more economical for the ailing space factory and its controlling computer to report the impending failure, and for the maintenance control center to request replacement by telecommunication. The factory computer, through its controlled robots, removes the ailing part, plugs in a replacement module, and sends the ailing module either back to the repair facility or into discard. This technology is already proven; the Soviets have accomplished very much the same thing with their Luna-return capsules. For some aspects of space factory repair and maintenance, this sort of completely automatic repair and replacement activity may be far more economical. For example, a very high-temperature space factory located in inframercurial orbit may require human life-support systems that are far too expensive in terms of the cost of repair and replacement.

The thought of operating and maintaining a producing space factory entirely by remote or automatic control should surprise no one who is aware of current (1975) state of the art in automation, computer technology, and remotely controlled devices.

It is possible even today for an individual to afford the necessary automatic and remote-control equipment to permit flying a small airplane entirely from the ground.

The current "hot" area in military flying is the RPV—remotely piloted vehicle. By radio link, pilots can now fly missions in drone aircraft with complete cockpit feedback, just as though they were sitting in the airplane themselves. Many new types of aircraft are being tested using this RPV technology.

Today's wide-bodied jet airliners are equipped with totally automatic flight systems. They can take off and land the airplane without a human hand on the controls.

The reader may think that I'm making a better point here for completely unmanned, automated space equipment, but this is not the case. Human beings no longer have to be right on site and in direct physical control; even with today's technology, it is possible for a human being to monitor, direct, and program a device from great distances, just as though he were there. But even though some of the functions that are automatic and repetitive have been taken over by automatic machinery, there is a human being "in the loop" exercising discretionary control override.

Thus, the majority of the people who run the space factories of the future will be highly trained technicians who have specialized in telecommunications, telecommand, automation, and cybernetics.

To some extent—at first only a skeleton force that later grows as space industry grows—these on-line space industry workers will be backed up by management teams familiar with more than just one small technical aspect of the space factory complex. Managers, coordinators, "big picture" people will be required in space industry just as they are on the ground. There is an old adage in industry that is always quoted in a humorous vein: "Every job needs a supervisor!" The principles of management—delegation of authority and responsibility accompanied by adequate controls—will be as important in space as they are on Earth. And the managers can't remain on Earth any more than all of them can remain in the head office in a distant city.

"You've got to be on the spot putting out the day-to-day brushfires. If you aren't, you just haven't got the full picture, and you can't make rational decisions that relate to the real world." This is in quotes because it is a paraphrase of the principles that several industrial managers have pounded into the author over the years.

The human institution of management and administration isn't the only area of human-to-human activity that will transplant itself into space. Other aspects will, too, because the Third Industrial Revolution isn't just transplanting physical plants into the space environment; it is the transplantation of industry, which we earlier defined as the sum total of many areas of human achievement.

For example, there will be conflict in space. It is as inevitable as death and taxes, which will also be there. Rules, codes, and laws will be required; someone will have to create them and write them down. Someone will have to interpret them. And someone will have to enforce them. As on Earth, these activities will fall upon the shoulders of the lawyers who are, in the final analysis, highly trained specialists in the resolution of conflicts between human beings.

This is not the sort of "space law" that has been pioneered by Andrew G. Haley, Nathan Schacter, William G. Hyman, *et al.* This is the sort of basic, operational human law of the same sort that was required in the American Colonies and in the American West. This is human law or common law translated into the space environment. It involves not only the basic problems of differences between whoever owns and operates the space factories, the sort of thing that probably will be handled in mahogany offices on Earth for quite some time yet, but adjudication of basic differences between human beings. It will be most interesting to watch what happens with true space law, whether the legal profession will grow to expand into the area of the Third Industrial Revolution or whether space law will develop on its own among the people who go out into space for space industry. One way or the other, it is bound to happen because rules and regulations governing behavior are basic requirements. And it requires people who are specialists in order to develop and interpret them.

We will also need human institutions devoted to enforcing the rules and regulations. We will take our police forces, security forces, and military/naval forces into space with us. Initially, they may well be like Pinkerton men. They may be the company guard force. In any event, they will be there to enforce, by physical force if necessary, the property rights of the corporations, individuals, or nations involved in space industry.

Yes, there will be property rights in space. Remember that hypothetical planetoid made of nickel-iron that is one mile in diameter and would supply our steel requirements for more than two centuries? That's valuable!

It doesn't do any good to lay claim to anything for your personal or collective use unless you are ready, willing, and able to defend your claim. This holds true of your life, your family, your home, your nation, your culture, your planet, or a planetoid in space. We are not yet altruists who have no property concepts; watch children at play. Better yet, in case you might believe that the concept of property is only a human trait, watch a litter of puppies.

The security groups of the Third Industrial Revolution will also be an interesting human institution to watch develop. They may begin as private groups. Or as mercenaries. Or as government military forces. Or as adjunct groups to the "risk industry."

The concept of insurance didn't start with Lloyd's of London and the merchant traders of England. It goes back to the Phoenicians and beyond. Basically, insurance is risk gambling. The underwriter bets with you that you are going to live long enough to pay enough premium to cover his costs, that your shipment is going to get through without damage, that your factory isn't going to burn down. And more. An underwriter works with a large statistical universe, betting on a sure thing.

Insurance is another human institution that will go into space. It will have to. Banks on Earth will require some sort of risk coverage for the advancement of capital required for facilities and working funds. Facilities will be insured. People will be insured. Cargoes will be insured in shipment. The underwriters may well remain on Earth, but they will have field agents out where the action is. It may be a different kind of insurance than is written today, but it will definitely be insurance.

And the insurance companies may well have to establish their own security groups to protect their risk.

Finally, we will have military groups in space . . . in spite of the UN treaties to the contrary. They may not be called military groups, but they will act in the same manner. They may take on the trappings of the present Coast Guard, a quasimilitary semipolice force. But they will have to be there to enforce the rules and regulations and to protect the investment of property and the lives of people.

About fifty centuries ago, the Phoenicians made one of the greatest inventions of all history and started a new human institution. This invention was money, and the institution is the bank. Before the invention of money, all trading was done by barter and all services and obligations were paid in kind. Money in the form of a token of universally recognizable value made from a substance easily tested and proven became a symbol for the original symbol of wealth: cattle. It became a symbolic means of keeping score of obligations. The next great invention of this sort was credit, the recognition that money is nothing more than a symbol. Credit is based upon trust of future payment. It is money in the future tense and has become negotiable all by itself.

A bank is not a storehouse for money any longer; it is a computerized scorekeeper.

The bank and other financial institutions are yet another human system that will participate in the Third Industrial Revolution. Initially, the roles of these institutions will be ground-based. Eventually, they will find their way into space, perhaps not in the form we know them today. They will have to go into space because people will be there and will require the services of keeping score.

Initially, the manpower of the Third Industrial Revolution will be transient; technicians will go into space for a "shift" of several weeks, returning then to the surface of Earth. Their living needs in space—food, water, air, clothing, entertainment, etc.—will probably be provided for them in space by their company, just as the living needs of the early astronauts in space were provided by NASA. However, the time will come when people go into space and stay in space for a long period of time, perhaps coming back to Earth only for medical needs or vacations; in this sort of a situation, some way will be required in space to keep score of consumables, to make certain that everyone gets exactly what is coming to him in the form of food, etc. "Keeping score" in this situation may at first start out with some form of life-support coupons or ration stamps. Whatever these tokens that entitle one to life-support materials are called, they are basically money. And they will be treated as money by the space workers because, as symbols promising the delivery of goods or services, they are negotiable items that can be traded or swapped.

If the firm issues these chits, they'll have to keep track of them, prevent them from being counterfeited, issue them, protect them, and honor them in return for goods or services. Somebody, therefore, has to count the beans and keep score. By any other definition, that's a bank.

Our culture is now far too complex to do without the concept of money and credit. We will not be able to do without them in the Third Industrial Revolution, either.

In due course of time as Earth-to-orbit transportation costs drop as the trend curves indicate they will, many other human institutions will follow people into space. They have followed colonists wherever they have gone on Earth. Initially, they will not exactly follow people into space; they will evolve in the space environment from the efforts of people who originally went into space to work in the space factories and, for one reason or another, struck off with little side-supporting businesses of their own in the classical marketing response: serving a perceived need or desire.

But why would people go into space to work as part of the Third Industrial Revolution?

For many reasons.

First, it will probably be one hell of a good job; it will pay well; it will have good fringe benefits. These are the same reasons why people will today go into the Arabian desert to run the oil wells or into the Alaskan Arctic to lay a pipeline. These space industry jobs will require the best our race has to offer—physically fit, mentally alert, highly intelligent, thoroughly trained people. Many people that fit this description are "loners"; they prefer the boondocks to the crowded, hectic rat race of the cities.

People will also become involved in space industry because it will be a transcendental experience of the sort that is already well understood by sailors, pilots, offshore oil men, forest rangers, naturalists, and other people who today prefer to work in a place other than Greater New York City. We have had only glimpses of the transcendental experience of space in contrast to that of flying, an occupation that has been well documented. These space glimpses were enough to move highly rational, factual, terse jet test pilots to stumbling words of poetry and to deeply religious personal revelations. To paraphrase John Gillespie McGee's "High Flight," several astronauts have already "put out my hand and touched the face of God." This transcendental experience may not move some of the early space industry pioneers; it's going to be a tough, hard, rigorous, deadly way to make a living. But I am willing to bet that should you catch one of these spaced workers after a couple of beers, you will find that it has touched him, too.

Thus, although the primary motivation for the Third Industrial Revolution will be monetary gain, it will eventually embrace nearly all human institutions, some of them developed painfully over long centuries of effort. When human beings go into space, they take with them more than their Earth-oriented life-support elements.

CHAPTER

13

Return to Eden

I DON'T KNOW where the saying came from that goes "Be careful what you ask for; you'll get it!" But it certainly fits in well with a generalized forecast for the year A.D. 2000 and beyond: "We will be able to do anything we want to do technically; we must be willing to do it, to pay for it, and to live with the consequences . . . *all* of the consequences."

Which in turn leads to 'the obvious question: "Very well. Suppose that we do manage to transfer most of our heavy, polluting industrial operation into the space environment in the next one hundred years. Suppose that we do cease plundering Planet Earth for raw materials and fossil fuels and instead find our materials and energy sources in space. Suppose that we do free ourselves from our earthly cage, our closed system. Suppose that we do open it up. What are the consequences? What problems will be solved by the Third Industrial Revolution, and what problems will it create? What will we be getting ourselves into?"

Again, we must stop and ask ourselves the questions "Who are we? Where did we come from?"

We have those answers. Our ancestors evolved during the past several glacial epochs into the most efficient hunters the planet has ever seen. Our ancestors were biologically attuned to the environment as it was after the Würm glacial age about fifty thousand years ago. By natural selection, the human body and, therefore, its mind as well are almost instinctively equipped for hunting.

Look at the hunting cultures that are left on Earth. What

characterizes the individual hunter? He requires space to move around in that is full of natural things—grass, trees, streams, rocks. He needs the presence of other animals. His job must be at times dangerous; at all times, his job must sharpen his wits, keep him on his toes, and challenge him. But he cannot work at all times; he must have the opportunity for exercise and recreation. Much of this recreation involves his need to seclude himself with a small group of friends that he can interact with in a face-to-face fashion. Although he has an outstanding ability to cooperate with others toward the achievement of a common goal, he needs to make decisions on his own regarding his contribution toward this goal.

We look out today upon our planet, our native home, and we see that we have generally not maintained an environment in which these characteristics can exist. To be attuned with the environment, we must therefore either create it artifically and take it somewhere else or re-create it here on Earth as best we can.

The Third Industrial Revolution will allow us to manufacture the goods we need without further destroying the Earth's biosphere. This, in turn, will permit us to begin restoring the biosphere to something resembling its condition fifty thousand years ago, a condition we grew up in from the evolutionary point of view.

But since there will be more than 6 billion of us living on the same land area that supported less than a million of our ancestors fifty thousand years ago, what we re-create will have to be new and unique and suited for the circumstances.

And since the number of people who will be required for the Third Industrial Revolution in space will be limited—"Everybody talking 'bout Heaven ain't goin' there!"—what in the world are 6 billion human beings going to do on Earth?

In the first place, there's a hell of a lot of work to do in the next century or more—probably more—cleaning up the mess, putting our home in order, and establishing the organizations that will help us keep it in order.

Our return to the Garden of Eden will find us keeping a

large number of industrial operations with us right here on the ground. This, again, is going to keep a lot of people busy. These "Eden industries" are going to be considerably different from those that we transfer into space. They will involve the final fabrication and assembly of products whose basic raw materials have come from space factories. These will be nonpolluting industries, assembly industries, permitting a considerable amount of individual craftsmanship. Remember that the Second Industrial Revolution will have matured so that nearly all repetitive operations are handled by automation and computers.

To some extent, there will be a resurgence of a form of "cottage industry," but cottage industry that is different from that replaced by the First Industrial Revolution. We see its bare beginnings today in the thousands upon thousands of small companies—machine shops, electronics companies, little single-product firms—that exist everywhere in America. We tend to think of industry as exemplified by the behemoths of du Pont, General Motors, and Union Carbide; we overlook the thousands of little Yankee machine shops stretched along the Connecticut shore, for example.

The Third Industrial Revolution will breed more of these small businesses. The opportunities for new products that can be made by these little firms will be vastly increased because of the new materials, devices, and products that will fall to Earth like manna from the space factories. Big business will operate the space factories; only they will have the capital clout to do it. But a big business is like a big truck; it has lots of inertia and cannot often respond quickly to new opportunities. In addition, big business is looking for big sales volume. This means that small business always has a place in the free enterprise system. Small business moves quickly. Small business will tackle the products for limited, specialized markets. Small business is willing to operate "tailor shops" for specialized "one-only" items.

The products of space industry will also require earthbound sales and distribution systems. With a space delivery system that can drop the product at New York, Agra, or Ulan Bator with only a slight variation in thrust vector, we will certainly be able to take a better cut at an equitable worldwide distribution system. The entire world is not like the U.S.A.; there are regions of unbalanced distribution. This sort of thing is going to cause real problems with customs duties and tariffs, by the way, because some factories are going to be owned and operated by socialistic, nationalized, state capitalism systems that are indeed national in nature and amenable to all current customs regulations; but others will be owned and operated by international free enterpise companies. Customs duties designed to protect local industries from undue foreign competition will eventually be eliminated because there won't be any local competition; shortages of raw materials and energy coupled with strict pollution laws will have eliminated the competition.

The products of the Third Industrial Revolution, whether they be from space factories or from the small plants of the benign industrial revolution, will be even more involved with high technology than today's technical products. They will need to be maintained and repaired. They will require the establishment of high-quality service and repair industries; some of these will be factory- or company-owned operations while others will be small privately owned shops. Yes, it's going to be just as tough in the future to get a gadget fixed!

The junk business is going to take on a new image. It will become the recycling industry, and it is already headed in that direction. We have learned that one man's garbage is another man's treasure. Even though we will have an extremely large source of raw materials in space, there will still be a requirement to reuse, refurbish, or recycle certain materials because of the high cost and/or scarcity.

A modern trend also indicates a continuing proliferation of service industries—people doing things for other people. Without the opening of Earth's closed system by the Third Industrial Revolution, service industries could evolve into the closed-system situation of everyone taking in one another's laundry. The influx of products and energy into the system from the Third Industrial Revolution in space provides the capital growth that will expand the service industries.

Communications—a word that takes in what we consider

"communications" today, plus transportation, the communication of people from place to place—will continue to grow. As Arthur C. Clarke has pointed out, with the communications satellite and the growing communications networks we are building, we are basically growing the nervous system of the collective human race. If we are to do the job that must be done with our lives and our planet in the next several centuries to survive this megacrisis that is upon us, we will need ever better communications. This is the only way to insure any sort of cooperation and coordination of activity. This is the only way to break down the barriers of mistrust and hostility that have grown from the basic neolithic philosophy.

This neolithic philosophy evolved in the first villages and cities built by people over five thousand years ago when our hunter ancestors settled down to the more certain life of farmers. Coon has paraphrased it as follows: "You stay in your village, and I will stay in mine. If your sheep come to eat my grass, I will kill you. However, I may need some of your grass for my own sheep. Anyone who makes us try to change our ways is a witch and we will kill him. Stay out of our village."

As Coon has also pointed out, this is the major stumbling block toward solving our planetary problems. Better communications is one aspect of the answer to this human problem. And the communications industry is going to continue to grow and grow because, when human beings run out of things to talk about, computers will still be talking with one another a thousand times as fast.

The other aspect of solving the problems brought about by this basic village philosophy will be another industry that will continue to grow on Earth during the Third Industrial Revolution: education.

Education is not only a desperate need of the human race right now because it is needed to overcome old, obsolete philosophies and concepts, but also because it is absolutely required if we are to increase and conserve the most vital of our human resources for the future: brainpower. Our ancestors became the world's most efficient hunters not because of strength of sinew or sharpness of claw, not because of fleetness of foot or speed of attack, but because of brainpower. Without it, we are the most helpless of animals. Without the accumulated and transmitted elements of brainpower— experience and methods—we are again protohumans at the complete mercy of the wind and cold and heat and fang. We cannot afford the luxury of discarding centuries of knowledge, or of reinventing the wheel, or of lethal trial-and-error learning. We cannot afford the animal solution of living with the problems; we must face and solve them as human beings have historically done. To do this, we must pass along to our progeny all that is good, noble, workable, and revealing from our history and our personal experiences. We must teach that which must be learned; experience has now shown us that a neophyte does not know what he needs to know.

Without education, the human race slips back fifty centuries in a single generation.

Six billion people must also eat, and feeding this many people is not going to be an easy thing to do. Agriculture and mariculture (farming the oceans, fishing, etc.) will continue to be industries that can exist only on Earth. Our planet is uniquely equipped to feed us. Although there has been some discussion of hydroponic farming on the Moon and in space for solving the food problems of Earth, it seems quite unlikely in the next hundred-year time period that we can consider this as a valid forecast. Can we feed 6 billion people? That's a subject for another book.

There will be a lot of things for the people of Earth to do in our return to Eden. But the step of moving the manufacture of our consumer goods into space to stop the destruction of our planetary biosphere by that industry will not solve the really big problems that are staring us in the face today. It will eliminate one problem and permit us to solve our remaining ones without entering the New Dark Ages of gloom, doom, famine, pestilence, death, and nuclear warfare that are the consequences of continuing closed-system Earth. It gives us hope that we may be able to attain for our children and grandchildren the same or better world of opportunity and plenty that we have built and enjoyed.

Some readers may feel that I have not spelled out in sufficient detail the possible time schedule of the Third Industrial Revolution. There are several reasons for this, the primary one being the admonishment of L. Sprague DeCamp: "It does not pay a prophet to be too specific." Industrial managers even today with modern techniques of data collection, computer correlation, and planning methodology have a hell of a time establishing a ten-year plan of any sort except in the area of long-range financing. There are too many variables. There are too many items that are beyond the control of each individual manager or each group of managers. And just when you think you've got it nailed down with all of the variables accounted for, you can be laid low by Murphy's Law: "Anything that can go wrong will go wrong."

This is why, even in the Third Industrial Revolution, we must not try to plan the future in great detail. We can paint it with a brush that gets broader and broader as the projection proceeds into the future. We can foresee the early years of the Third Industrial Revolution because they are already taking place and proceeding along a logical path of action.

The same holds true in any discussion of and plan for consequences of the Third Industrial Revolution.

The only way we can realistically forecast and prepare for the future is to discover the key points and problems . . . and then follow through on them in a manner that many futurists and planners might call opportunistic.

There is much to be said for opportunism carried out against a generalized long-range overview that has identified the problems. We cannot control every facet of life. We cannot outguess the future. We cannot anticipate the revelations of serendipity.

This book and others like it that attempt to solve problems rather than wallow in them can only point out a potential course of action and review the known factors involved. It is up to individual people and groups of people to do something, to have the belief in a better future that gets them moving, to be willing to risk life and money and energy in the calculated hope of getting more out than they've put in.

The Third Industrial Revolution solves only one problem of the future and leaves us free to tackle some of the others.

We will have to apply the same sort of opportunistic long-range general planning to our return to Eden. Getting rid of our polluting, planet-eating industry and putting it in a better environment does not solve the other problems that will still be there, big as life and twice as ugly, as the Third Industrial Revolution begins to supply us from space. These big problem areas that we're going to have to keep generally in mind are:

1. *Energy Production.* Unless we develop a pretty good "bucket" to transport back to the ground the virtually limitless energy available in space, we will have to find other energy sources here on Earth that will continue to support social institutions of increasing size and complexity. These institutions are absolutely necessary if we are to forge the difficult bonds of planetary unity required by the future.

2. *Education.* Over 30 percent of the Earth's population cannot read or write their native language. Eight hundred million minds are unable to communicate except by the spoken word or pictures. To cope with local problems that have a worldwide effect when combined, and to prevent these problems from becoming runaway catastrophes in a world of increasing knowledge and technology, the human race is going to have to do a better job of education in order to achieve the full utilization of brainpower. It is of utmost importance to us all, Third Industrial Revolution notwithstanding, to educate every person up the point where he can fully utilize his own individual talents.

3. *Population Control.* This is perhaps the biggest problem of all. Our biological heritage as hunters demands that we completely fill what appears to be an ecological niche, that we breed up to the point of numbers where natural limitations restrict that breeding. The public health official and the sewage engineer so completely upset the previous ecological human balance—quite unintentionally and with the highest of

all motives—that we may not recover from it as a species before we starve ourselves to the natural population level that Planet Earth can support. Malthus may yet be right, but we still have the opportunity to prove him wrong through education. The industrialized nations may at the current time be voluntarily limiting their population growth, but try to convince a Third World farmer that he should not sire as many offspring as he is able for two reasons: (1) he needs more help around the farm, and (2) 50 percent of his children will die before reaching the age of five. The interim solution to this problem that might be resorted to while a worldwide educational program is getting under way can best be summed up by another statement by the incredible Robert A. Heinlein: "What the world really needs is a safe, oral contraceptive combined with a habit-forming soft drink."

The Garden of Eden—Earth—that we slowly return to as the Third Industrial Revolution reaches its own climax of maturity a century hence will therefore have many of our current problems intact, and they will seem even more important because of the fact that we were able to solve one of our problems by moving our industrial base into space. Yes, it will be a garden planet again with work for all and plenty of problems yet to solve for the future.

But there are solutions for these problems of the now-and-future, just as the Third Industrial Revolution promises to be a solution to one of the pressing problems of our twentieth century.

There are always solutions, even to the most impossible problems, because the human brain working with logic and a knowledge of the way the universe works can organize millions of other human brains to apply the knowledge to solve the problems. This is not a matter of blind faith, but a knowledge of how we got to where we are and, through the same procedures, how we will probably prevail. We have yet to discover our real limitations. We have yet to run up against any problem with absolutely no conceivable solution. When we do, the universe will let us know in very certain fashion. But it hasn't happened yet.

CHAPTER
14
The Action

TALK IS CHEAP, and words on paper are just that. The Third Industrial Revolution will not take place because I or anyone else writes about it. Somebody is going to have to do something about it. People are going to have to make it happen.

These people are managers, executives, entrepreneurs, and risk-takers.

These people always want to know not only what can be done, but also what they can do. They ask, "What action is required?"

It's early in the dawn of the Third Industrial Revolution. In typical, historical fashion, government is footing the bill for exploration until things begin to snowball. A glance at "Commerce Business Daily," published by the U.S. Department of Commerce and listing all requests for proposals and bids of the government reveals the following on January 16, 1975:

> Development of containerless process for preparation of tungsten—8-1-5-58-00130-AD13-C, 9 Jan 75, negotiations will be conducted only with General Electric Company Valley Forge Space Center, P. O. Box 8555, Philadelphia PA 19101 (PO13), Procurement Office, NASA George C. Marshall Space Flight Center AL 35812.
> Convection in space processing ASTP science demonstrations—One job—Delivery FOB Huntsville AL DCN 1-5-59-51040-AD13-B, negotiations will be conducted with Lockheed Missiles and Space Company, Huntsville

Research and Engineering Center, Huntsville AL—
Copies of RFQ not available (PO13), Procurement Office
NASA George C. Marshall Space Flight Center AL 35812.

The action has started. Soon it will be time to grab a bucket
because it will be raining soup.

The first step that must be taken in order to participate in
the Third Industrial Revolution is the same as one would
take if he were to move himself or his company into any new
area of marketing, product, or process: Collect data. This is a
low-cost operation to start with requiring the time and effort
of one or two people who will dig around and find out what
is going on. Don't assign it to your director of research if you
cannot do it yourself, unless your director of research is a
very unusual, wide-ranging encyclopedic synthesist. If he
doesn't know what that is, he shouldn't be given the job! Assign it only to a marketing man if he is young and has a technical background.

NASA is probably the best source of information at this
time. And the fountainhead of information seems to be the
George C. Marshall Space Flight Center in Huntsville, Alabama.

The next step is to study the information and identify potential areas for your organization to become involved in.
What potential space products will provide competition to
your current products in the next ten years, fifteen years,
twenty-five years? What potential space processes will result
in improvements of your current products? What potential
space products will provide a logical expansion of your current product lines?

This sort of thing may have to be done by somebody that
you will have to bring into your company from outside,
somebody who is not fettered with the corporate optimism
and sales publicity of your present products. This always exists, even in a small company. It is the NIH Factor—"Not Invented Here." It is the "we-can-do-no-wrong" syndrome. It is
the company shibboleth that "we are the best and our com-
petitors are a bunch of nice but misguided guys who are

turning out a definitely inferior product." The sales and marketing departments are usually full of this; they cannot function at full power without it. It may have trickled down into the production, engineering, and research and development departments, too, which spells real trouble because they really don't know what the competition is doing at all and have never really analyzed the competitive products.

By all means, do not obtain the opinions of eminent men in their fields, whether or not these men are on the board of directors. There are exceptions to this, of course, but Arthur C. Clarke's First Law is generally valid: "When an eminent scientist states that something is possible, he is most assuredly right; when he states that something is impossible, he is most assuredly wrong." Competence alone is not enough, nor is imagination alone enough. According to Dr. Robert M. Wood of McDonnell Douglas Astronautics Company, what is needed is "imaginative competence," the man who reads *Fate* magazine as well as *Product Engineering*, the person who reads science fiction, can run a lathe, and knows partial differential equations.

Sooner or later, when you go beyond these early exploratory steps, you are going to have to build a team of proto-space engineers, a development group that will, you hope—if you've done things right—emerge as your operating space industry group a decade hence. Staff it now with mechanical engineers and physical scientists and electronics men and computer specialists who are hot-rodders, sports-car drivers, ham radio operators, and other forms of technical hobbyists. I have never met a person who was a professional as well as an ardent technical amateur who was afraid of the future or fearful of change or who really worshiped sacred cows of science or technology.

In spite of the fact that it may be a little bit early to conduct market studies for space products, the pioneer work in this field has already been done by David Keller, manager of advanced programs of General Electric's Valley Forge Space Center. Keller presented his marketing data as part of his testimony before the Senate Aeronautical and Space Sciences

Committee on October 31, 1973; Senator Barry Goldwater presents the results of this marketing study in detail in the Introduction to this work. It is also listed in greater detail in the *Congressional Record,* Vol. 120, No. 47, dated April 3, 1974.

It is possible to do a market study based on your findings, and it should be done to pinpoint factors that you might have otherwise missed. This market study will get your space industry development group off to a proper start in the right direction: planning for products that will sell and produce a return on the investment at the earliest possible time. Without a market study to guide future work, you run the risk of dibbling and dabbling around in interesting little projects and problems that may be dear to your hearts but too long-range to keep your stockholders happy.

Market studies will also give you some feeling for the time frame you will be working within and, therefore, the budgetary level you must live with.

If you identify space-processing areas of interest, and if your marketing studies indicate that you will be able to achieve a return on investment within whatever period of time you determine, under your circumstances, to be "reasonable," it's time to start actual testing, process development, and product investigations. For the next decade until private shuttles become available, this sort of thing will have to be done in cooperation with NASA. Your programs will need zero-g jet flights, space on the plethora of sounding rockets that NASA will launch for space-processing experiments, and eventually space aboard the Shuttle and *Spacelab* in 1980 and thereafter.

The Third Industrial Revolution is still young enough to permit one to hedge his bets while still throwing dice and betting on the come. It is young enough that you can get into the action for a modest piece of change, a minimal investment of time and effort to find out what can be done. It is possible to cash in or change direction without having to commit for a fullfledged space-processing facility in orbit.

But it won't be that way for long.

A most important factor is to keep abreast of developments continually because they will come faster and faster. If the time doesn't look ripe yet, if the combination of many factors doesn't "feel right," if your gut feeling tells you it is still too early, keep on top of it. It will change rapidly.

Much of what will be done in the coming decade will appear to be a minimal amount of action. But don't be fooled; it is likely to be a triggering mechanism.

A triggering mechanism is just what the term implies: a device or system that will, by the application of a very small amount of energy, release a disproportionately large amount of energy. There are many examples of this: ink on paper, a finger on a switch, one neutron in a 137.3-pound sphere of uncompressed Uranium-235, and the straw that broke the camel's back.

Much of what goes on in space processing and the Third Industrial Revolution in the coming decade will be triggering mechanisms for what will follow.

The first billionaire space moguls are now alive.

There is nothing that guarantees that they will be Americans; the Second Industrial Revolution peaked on the other side of an ocean from where the First Industrial Revolution began. There is only one thing that appears to be certain: The Third Industrial Revolution will occur in the same sort of culture that fostered the first two. That sort of culture can be characterized by a maximum amount of freedom for the individual and the ability for the individual to better his lot by taking the maximum risk. This is the sort of culture that has characterized the progress of human development through the centuries, which is why the statement can be made in such a positive manner. It is also characterized by the ability of people to freely form whatever social organizations and institutions are required for getting on with the job, maximizing the return, and doing all of this with the minimum of conflict and dislocation.

Regardless of who the space moguls are, they and their cohorts will have drive and ambition, a positive philosophy of life, and the ability to work hard toward a plausible goal.

These traits have characterized Americans for more than a century, but there is no insurance that these traits are still strong enough. We know very little yet about what makes a culture mature and grow senile; we do know what makes it grow.

There are only three nationalities that appear today to have the vitality, the spare effort, and the competitive drive available to spawn the space moguls: the United States of America, Japan, and the Federal Republic of Germany.

But how about the Union of Soviet Socialist Republics? How about the People's Republic of China? Perhaps, but their efforts today are almost completely occupied with their internal development, the creation of a suitable industrial base, and a military activity to protect them from each other along a common border. The impressive Soviet space program appears, on closer inspection, to have been based on a refinement of 1945 rocket technology spurred by the life force of two men: Nikita Krushchev and Sergei Pavlovitch Korolyev, the Soviet von Braun, who died in 1966. The cooperative space efforts of the 1970s have shown us that Soviet space technology is about ten years behind that of the U.S.A. and is stretched to the limit. In addition, the U.S.S.R. still suffers from a "short blanket" economy: When the shoulders get the blanket, the feet get cold. By the turn of the century, we might expect to see the U.S.S.R. involved more deeply in space industry, but hardly to the level of the three countries listed above. There are strong doubts that the People's Republic of China will be a member of the club by that time.

But should we also list Brazil? Australia? The Republic of South Africa? Iran? These nations, among others, show encouraging signs of attaining the characteristics needed for a capital-intensive expansion such as space industry: stable political climate amenable to the sort of entrepreneurial drive demanded, a growing technical and educational base, adequate expansion of the capital base, and the driving, fire-in-the-belly guts of the people.

Perhaps because of our growing "nervous system" of in-

stant communication, we will suddenly discover that the
Third Industrial Revolution knows no nation, that it has itself spawned new supranational organizations and institutions. There are very few political boundaries that can be
seen from space, and orbital space can have no national political boundaries. It is patently impractical for a nation to claim
the Moon, a planet of almost 15 million square miles of surface area, and proceed to defend it with space cruisers,
Moon-based laser guns, and Moon-launched antispacecraft
missiles; to set up customs and duties and tariff walls; to coin
Moon money and issue valuta.

The Third Industrial Revolution may well make today's
nations seem as quaint and primitive as the city states of
Greece are to us today.

The space moguls may be as supranational as today's mul-
tinational corporations and multimillionaires.

And with the Third Industrial Revolution, they will be
completing a process that started over ten thousand years
ago at the beginning of the Neolithic Age. Before that time,
people lived in small, nomadic hunting groups in a tribal culture where most children died, where the tribe might perish
completely, where the people went hungry most of the time,
and where a person was old and ready for death at age thirty.
The big change started when a tribe looked down into a valley and saw a cluster of dwellings surrounded by tilled fields;
the Neolithic Age had begun. This was a barbarian culture
based on an individual moral code, based, in turn, on the
concept of property. It grew into a peasant culture, an
agricultural way of life where everyone had a little bit of
something, but nobody had very much of anything. Again,
half the children died before reaching maturity, and a person was old at thirty-five. The circle of huts grew into a village, which soon started to trade with the neighboring village
in spite of considerable violence that existed between them.
The village on such a trade route grew into a city, especially
if it was located in a place that could easily be fortified and
defended, thus permitting the local rulers to offer protection
and charge a toll for passing unmolested. The city bred a

new sort of people: civilized people who learned a new way of cooperating, getting along, and living in large groups.

As all this has been taking place, the ills have grown, the problems have multiplied. Only the industrial revolutions have broken a weary sequence of invade, conquer, rule, be assimilated, and be invaded. Only the industrial revolutions and the increased trade thus produced have broken the cycle of march and countermarch of armies, looting and raping as they came and went. Only the industrial revolutions have broken the age-old philosophy of taking by force from those who have and replaced it with a philosophy of make it, instead of take it; trade for it, instead of conquer it; create customers, instead of take slaves.

The dawning of the Third Industrial Revolution in space holds the real promise of completing this first major cultural, moral, ethical, and philosophic shift to take place since the beginnings of the Neolithic Age one hundred centuries ago.

The Third Industrial Revolution holds the promise also of righting many of the wrongs that áre consequences of the first two industrial revolutions, the industrial shifts that were necessary to create the wherewithal for the Third to take place.

The Third Industrial Revolution is the action of the future because it appears to be consistent with the long-term trends in human history and development. It is a logical and feasible direction in which to move, and the movement has already started even without a philosophical base to initiate and sustain it to date.

The Third Industrial Revolution appears to be moving along right on schedule and just in the nick of time. If it doesn't prove out, if it doesn't happen, Meadows and his colleagues are right. Malthus will also be proven correct. And this is totally inconsistent with human history and development. If the Third Industrial Revolution is not a realistic forecast, perhaps it is the fate of all intelligent, self-aware species in the universe to blaze like an exploding star for one brief instant of climactic glory before sinking back into a final nuclear dark age.

I, for one, prefer to believe that there is more to the human race, that we have come this far and will not be daunted or denied.

Ralph Waldo Emerson wrote in *Politics*: "We think our civilization near its meridian, but we are yet only at the cock-crowing and the morning star."

Of all the revolutions of yesterday and today, the Third Industrial Revolution is going to be the most fascinating and satisfying of all revolutions to take part in.

Appendix

INDUSTRIAL FIRMS INVOLVED IN
SPACE PROCESSING RESEARCH (1975)

American Optical Company
Grumman Corporation
Rockwell International
Martin Marietta Corporation

General Dynamics, Inc.
Bendix Corporation
Chrysler Corporation
General Motors Corporation
National Lead Company
Reynolds Metals Company
Western Electric
Bell Laboratories
Teledyne/Brown
Monsanto

General Electric Company
Tyco Labs
Revere Copper & Brass
Westinghouse Electric
    Corporation
Lockheed Company
Boeing
E. I. du Pont de Nemours
International Business Machines
Owens-Illinois
Union Carbide Corporation
Texas Instruments, Inc.
United Aircraft Corp.
Thompson Ramo Wooldridge

Bibliography

Bono, Philip, and Gatland, Kenneth. *Frontiers of Space*. London: Blandford Press, 1969.

Cole, Dandridge M. *Beyond Tomorrow*. Amherst, Wis.: Amherst Press, 1965.

Coon, Carleton S. *The Story of Man*, 3rd ed. New York: Alfred A. Knopf, 1969.

Ehricke, Krafft A. "Exploration of the Solar System and of Interstellar Space." *Annals of the New York Academy of Sciences*, Vol. 163 (1969).

———"In-Depth Exploration of the Solar System and Its Utilization for the Benefit of Earth." *Annals of the New York Academy of Sciences,* Vol. 187 (1972).

Haley, Andrew G. *Space Law & Government*. New York: Appleton-Century-Crofts, 1963.

Libassi, Paul T. "Space to Grow." *The Sciences*, Vol. 14, No. 6 (July/August, 1974).

Meadows, D. H., Meadows, D. L., Randers, J., and Behrens, W. W. *The Limits to Growth*. New York: Universe Books, 1972.

O'Neill, Gerard K. "The Colonization of Space," *Princeton Alumni Weekly* (November 12, 1974).

Ruzic, Neil P. *Where the Winds Sleep*. New York: Doubleday & Co., 1970.

Stine, G. Harry, "The Third Industrial Revolution, Part I." *Analog*, Vol. XC, No. 5 (January, 1973).

———"The Third Industrial Revolution, Part II." *Analog*, Vol. XC, No. 6 (February, 1973).

———"The Third Industrial Revolution: The Exploitation of the

Space Environment," *Spaceflight*, Vol. 16, No. 9 (September, 1974).

WILLIAMS, J. R. *The Bull of the Woods and Out Our Way.* New York: Charles Scribner's Sons, 1952.

WOOD, DR. ROBERT M. "Giant Discoveries of Future Science." *Virginia Journal of Science*, Vol. 21, No. 4 (1970).

————The *Congressional Record*, Vol. 120, No. 47 (April 3, 1974).

————"Free Atoms: A Whole New Basic Chemistry." *Science News*, Vol. 100 (December 11, 1971).

————"Marshall Studies Materials Processing in Space." NASA George C. Marshall Space Flight Center Release No. 74-47 (April 4, 1974).

————"Proceedings of the Third Space Processing Symposium, Skylab Results." NASA M-74-5 (1974).

————"Processing Vaccines in Space." NASA Release No. 74-72 (April 3, 1974).

————"Soviet Space Research." Moscow, 1972.

————"Space Processing and Manufacturing." NASA ME-69-1 (1969).

————"Space Promises Better Materials Processing." NASA George C. Marshall Space Flight Center Release No. 74-33 (March 8, 1974).

————"Unique Manufacturing Processing in Space Environment." NASA ME-70-1 (1970).

Index